DRAWING
for Engineering

Paul Smith

JUTA
AND COMPANY LTD

First pubished 1999
Reprinted 2011
Reprinted 2013
Reprinted 2014 (Fourth reprint)
Reprinted 2016
ISBN 078 0702 144066

P.O. Box 14373, Lansdowne, 7779, Cape Town, South Africa

Editors: Marion Boddy-Evans, Karen Press
Illustrators: Dennis Bagnall, Renato Balona, John Small
Book design and typesetting: JK Type & Graphic cc, Roodepoort
Cover design: Eugéne Badenhorst
Print and bound by Mega Digital (Pty) Ltd

Acknowledgement:

The authors and publishers wish to thank the following persons and institutions for their invaluable contribution to the development of this publication:

- The United States Agency for International Development (USAID), for funding the project. This Materials Development Project formed part of the Tertiary Education Linkages Project (TELP) which focused on capacity building at Historically Disadvantaged Technikons through the establishment of linkages with universities in the United States of America.
- Contributors and moderators from the following South African institutions: Mangosuthu Technikon, ML Sultan Technikon, Peninsula Technikon, Technikon Eastern Cape, Technikon Northern Gauteng, Technikon Southern Africa.
- Contributors and moderators from the United States University Consortium comprising Howard University, Massachusetts Institute of Technology, Clark Atlanta University, North Carolina A & T University.

We thank the following institutions for permission to reproduce material:

- Delmar Publishers, Albany, USA, for permission to reproduce figure 33.2 of Olivo, et al., *Basic Blueprint Reading and Sketching (5th edition)* on page 69.
- The South African Bureau of Standards for figures 20, 23, 30, 41, 42, 43, 56, 57, 58 and 61 on pages 29, 31, 39, 52, 54 and 66, and the Table of Abbreviations in Appendix II from *SABS 0111: Part 1 – 1990, SA Standard Code of Practice for Engineering Drawing (2nd revision).*
- Heinemann Publishers for permission to reproduce figures 112 and 113 from: Moolman, C.L. 1980. *Engineering Drawing NTS3* on pages 177 and 178.
- Metric Publications for permission to reproduce figures 1, 2, 3, 4 on page 15; 1, 8 on page 53; 1 on page 57; 1, 4, 9 on page 61; and 1 on page 71 of Elliot, V. *Technical Drawing N2.*

Table of contents

Preface

This book is presented in the conventional order in which the various topics are generally taught at technikons. The scope of the unit is such that it will form the foundation of Engineering Drawings (Mechanical and Chemical) at a tertiary level.

This book is specifically designed to cover the requirements of ISO standards and also the SABS standards (Code of Practice for Engineering Drawing SABS 0111) and should, therefore, give the learner technician an idea of the fundamentals of draughtsmanship.

The presentation of the topics in this book is such that students should be able to handle most of the work with a minumum of supervision and for this reason it is recommended for distance education as well.

It is necessary to urge that students get plenty of practice in free-hand sketching. To be able to make neat, well-proportioned freehand sketches is as important as the ability to make drawings at a drawing-board using a T-square and drawing instruments. In the Appendix, all the abbreviations are shown used on Engineering Drawing.

This is an ACTIVITY icon. If you see this icon then you will know that it is time to DO something! The activities help you to understand the subject. Feel free to do them with a friend or a group of friends. Whatever you do, just do them! The solutions to the activities are given at the end of each unit.

This is a DEFINITION icon. Read the definitions carefully because the details are important.

This TAKE NOTE icon appears alongside all the extremely important information.

Unit 1

Drawing equipment

Study objectives

After studying this unit, you should be able to:

- ◇ **Identify** and **name** the different types of instruments
- ◇ **Apply** the relevant drawing equipment correctly for drawing the different types of drawings or constructions
- ◇ **Describe** the 18 essential dos and don'ts of neatness in drawing and **explain** why neatness habits are important.

1 Introduction

If you read through the above study objectives, you should understand the aim of this particular unit, which is to enable you to **name and use drawing equipment correctly**.

Naturally you will need to be able to identify and accurately name the tools you will be using to start and finish a working drawing.

2 Drawing equipment

You do not need much equipment for this course. As a beginner you would be wise to seek advice from an **experienced** draughtsperson, lecturer or even a student before purchasing expensive sets of instruments. The reason is that these sets generally contain more items than you initially need. Any learner working through this drawing course **must do a great deal of practical homework**. It is important that you use your equipment over and over again correctly until you can draw neatly and efficiently.

This course you are doing on drawing is definitely a **practical** one – a 'doing' course. This means that it is only by constantly **practising** the techniques, using the correct instruments in the right order, that you will become effective and efficient as a draughtsperson. Always make sure that you are using the correct techniques.

We will now discuss the **12** pieces of drawing equipment generally used for this course.

2.1 Drawing paper and its application

Sheets of drawing paper are available in various **standard sizes** ranging from A0 (the largest) to A6 (the smallest). The measurements of the different sheet sizes are as follows:

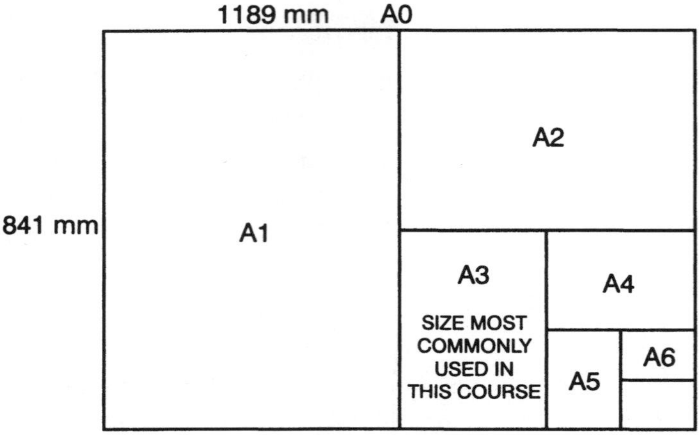

A0 (841 mm × 1 189 mm)
Al (594 mm × 841 mm)
A2 (420 mm × 594 mm)
A3 (297 mm × 420 mm)
A4 (210 mm × 297 mm)
A5 (149 mm × 210 mm)
A6 (105 mm × 149 mm)

Fig 1.1

It is interesting that the area sheet A0, which measures 1 189 mm × 841 mm, is equal to one square metre (1 m^2).

Did you notice that when sheets from A0 to A5 are folded in half or just halved, the next smallest standard size is produced?

Using drawing paper

A0 and Al sizes are usually used for drawings for construction projects. The choice of sheet size depends on the size of the drawing to be done.

For this course, we recommend that you use A3 drawing paper because in the early stages of developing drawing skills, A3 is easier to work with. If the A3 paper weighs 120 g/m^2, you can use both sides as it is then thick enough.

If you work lightly you can practise on one sheet of paper over and over again if you feel you should do so. All you need to do is rub out the lines. In this way you can cut down on expenses by saving paper.

2.2 Drawing board and its application

The drawing board is usually a timber board with a smooth and absolutely flat surface. It has a straight working edge on the left-hand side which is sometimes fitted with a metal or a plastic shield. The working edge is for the T-square (see section 2.3) to slide on, and there is a true right angle between the T-square and this edge (see fig 1.2).

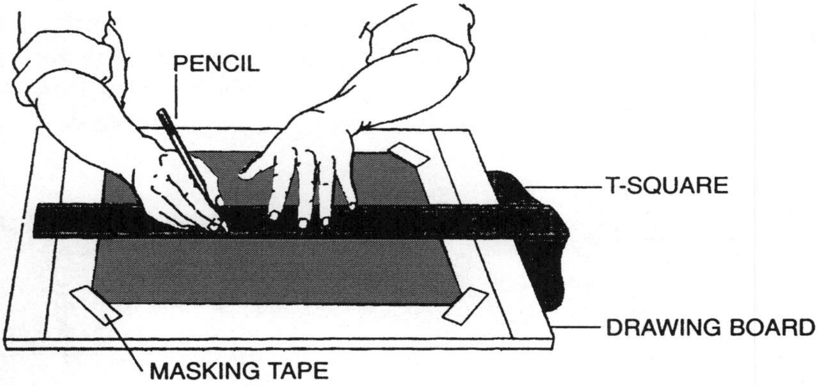

Fig 1.2 An A3-size drawing board

Using the drawing board

Drawing boards are available in different sizes, depending on the size of drawing sheet a draughtsperson wants to use. For this course, we recommend that you use an **A3 drawing board** because it is smaller and easier to carry than an A2 drawing board. These study units contain various exercises which will require you to do drawings yourself. You can do these drawings on sheets of paper which you must fasten to the surface of your drawing board so that you have a smooth, even surface on which to work.

2.3 T-square

You should use a T-square that is suitable for use with an A3 drawing board, as shown in fig 1.3.

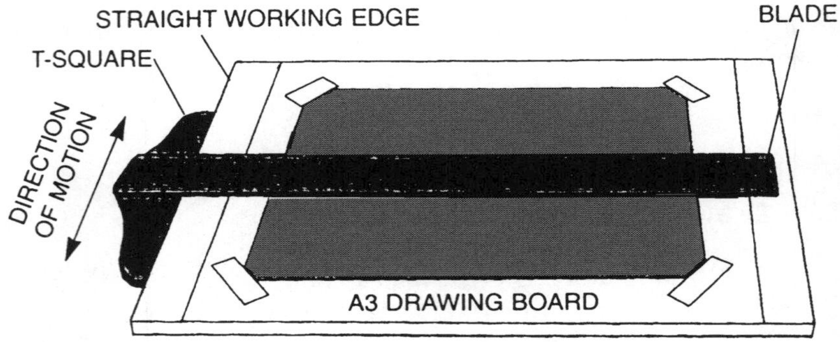

Fig 1.3

Using the T-square with the drawing board

The T-square is used mainly for **drawing horizontal lines** and for **guiding set squares** when drawing vertical and inclined (sloping) lines. It is moved by sliding the guiding edge of the head along the left edge of the drawing board until the blade is in the required position (see fig 1.3 above and 1.4).

Fig 1.4 Apply pressure to the side of the T-square. Keep pressing towards the right to keep the T-square tightly secured to the edge of the drawing board

---

Activity 1

Look at the drawing below of a scene from our everyday environment. Try to fill in the blank spaces in the sentences that follow. (Write your answers in your roughwork book.)

Note: the road is horizontal

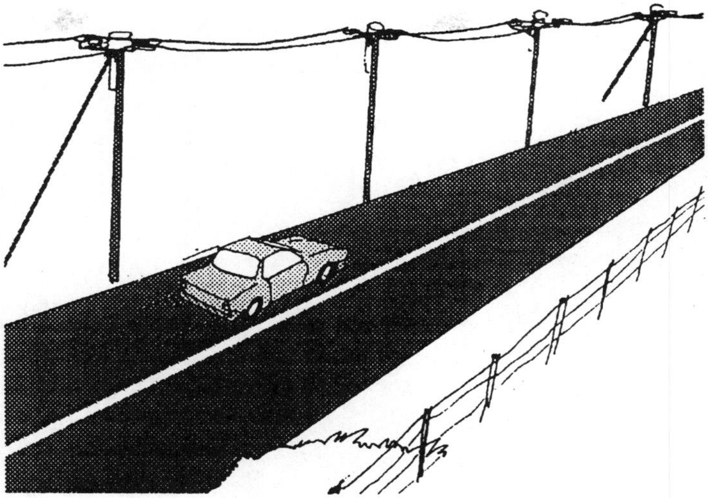

Fig 1.5 An everyday scene

1 The telephone poles are _____ to the surface of the road.
2 Therefore we can say that the telephone poles are _____ .
3 The telephone lines are h_____ and more or less p_____ to each other.

Remember that you can find the answers at the end of the unit. You must compare your own answers with the correct answers. In this way you can monitor your understanding of the learning material. As we explained to you in the introductory letter, you are encouraged to evaluate your own progress and assess your understanding of concepts.

2.4 *Set squares or triangles and their application*

You will need set squares with angles of 45° and 30° / 60°. The 45° and 30° / 60° set squares are commonly used for **ordinary** drawing practices. The length of the shortest sides of the set squares must be approximately 200 mm. Chips and nicks on the edge of a set square makes it useless, so check the edges regularly as shown in fig 1.7.

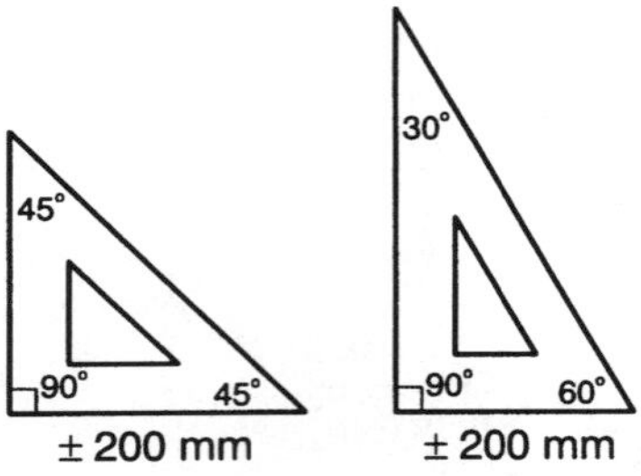

Fig 1.6 Set squares

Fig 1.7 Testing set squares for chips (nicks)

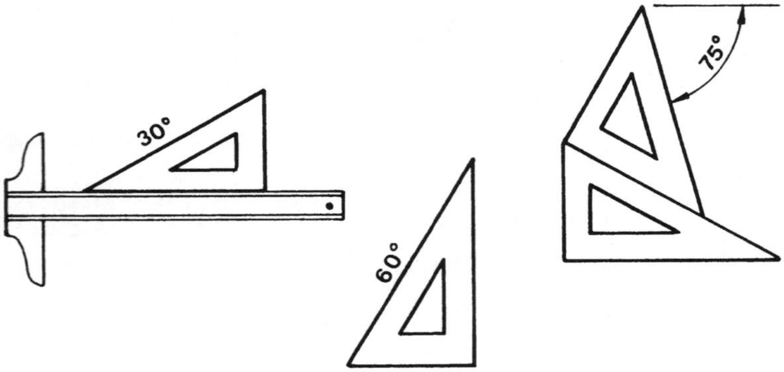

Fig 1.8 Drawing different angles using the set squares

2.5 Masking tape

Masking tape is used to fix the drawing paper securely to the drawing board (see fig 1.2). Unlike sellotape, it does not damage paper.

Activity 2

Practise drawing different angles with the set squares and T-square by copying the circle and its parts, as shown in fig 1.9, on an A4 or A3 sheet. The radius must be 7 cm. Use only your compass to draw the circle, and then your set squares and T-square to measure and draw the angles (see fig 1.8) Vertical lines are always drawn away from you. Other lines can however be drawn in either direction, depending on whether you are right- or left-handed. You always **draw** (pull) a line and never push the pencil, which may cause the point to break.

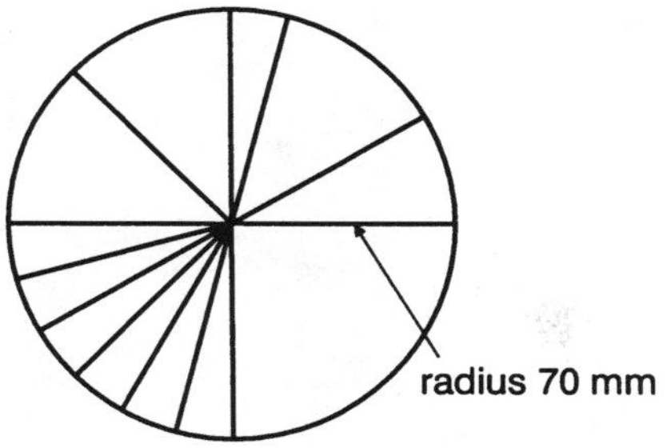

Fig 1.9 Drawing lines at different angles

2.6 Pencils and their application

Graphite pencils are usually made in 17 degrees of hardness, as shown in the table below, the most common are 2H, H and HB.

Table 1.1

Degrees of hardness			
6B	softest and blackest	H	medium hard
5B	extremely soft	2H	hard
4B	extra soft	3H	hard, plus
3B	very soft	4H	very hard
2B	soft, plus	5H	extra hard
B	soft	6H	extra hard, plus
HB	medium soft	7H	extremely hard
F	intermediate, between soft and hard	8H	extremely hard, plus
		9H	hardest

The pencil point

When using an ordinary pencil, the type of point you use is very important. Many learners prefer to use a **conical** point, while others find a **wedge** point more suitable for **straight-line** work as it requires less sharpening and makes a denser line.

When sharpening a pencil, you should cut the wood away (on the unlettered end) with a knife or a pencil sharpener. About 5 mm of the lead should be exposed and shaped to the desired type of point – **conical** or **wedge** (see fig 1.10). The hardness of the lead (2H, H, etc.) is indicated on the lettered end of the pencil.

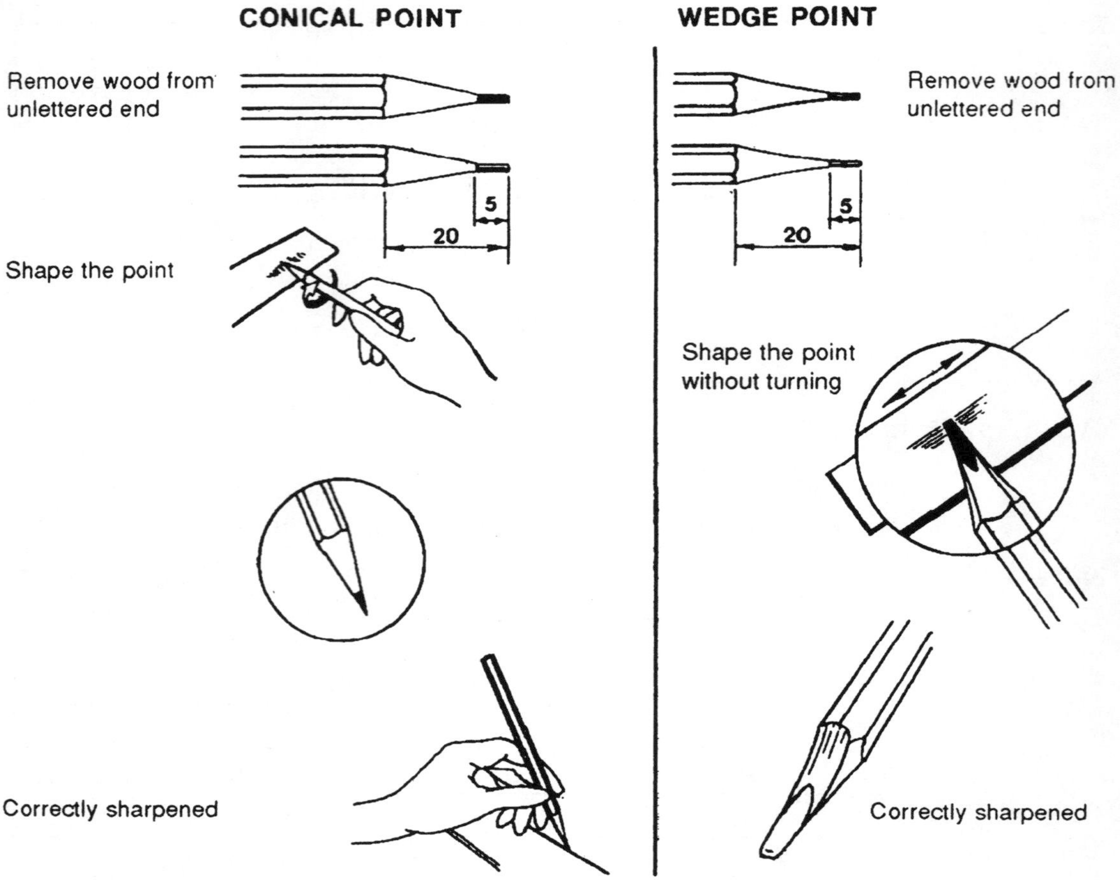

Fig 1.10

Instead of an ordinary pencil, you can use a **clutch** pencil.

Fig 1.11 A clutch pencil

We recommend that you rather use a clutch pencil for the following reasons:

- ◊ You can draw **better quality lines**
- ◊ Using clutch pencils **saves time**
- ◊ Using clutch pencils improves the **general neatness** of a drawing since no messy sharpening is required

The leads for a clutch pencil come in a separate container and the hardness of the leads or points is clearly marked on the container. Table 1.2 shows the different leads and their uses.

Table 1.2

Lead thickness	*Hardness of lead*	*Recommended use*
0,3 mm	2H	faint lines, e.g. construction lines, projection lines and dimension lines (which we will be discussing in Unit 2)
0,5 mm	H	outlines
0,7 mm	2B	outlines on construction drawings

Using a pencil to draw lines

When drawing a line, you should incline or tilt the pencil slightly (about 60° in the direction in which the line is being drawn).

The pencil should be **pulled**, never **pushed**, at the same inclination for the full length of the line. We always **draw** lines **away from us**. If you **rotate** the pencil slowly between your fingers as the line is drawn, you will be able to maintain a **symmetrical** pencil point and will be guaranteed a straight, uniform line.

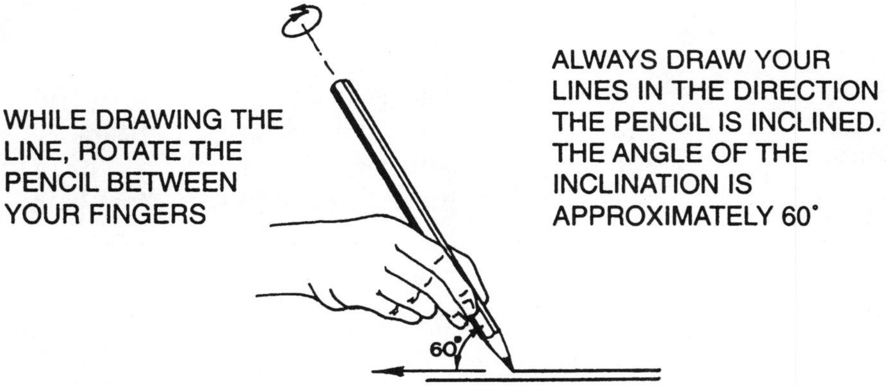

Fig 1.12 Drawing a line

A symmetrical pencil point means that the point will remain shaped and not lose its pointed shape so quickly.

Activity 3

1 What shape will the pencil point be if you use the rotation technique explained above?—

2 Give a reason for your answer.

2.7 Eraser (rubber)

Fig 1.13

You need a **good quality** eraser when doing a drawing. When you are finished with a drawing there will always be a lot of unnecessary lines to be erased (rubbed out). Smudging or clouding from using a dirty or poor quality eraser must be avoided at all costs. Always rub out unwanted lines and tag ends by stroking the eraser in **one direction** only.

Activity 4

Can you think of **two** reasons why you must always use the technique of stroking in one direction only when you rub out? Write down what you think.

2.8 Erasing shield

The erasing shield is used together with the eraser. When you rub out unnecessary lines or work, position the shield so that the appropriate opening is over the line to be rubbed out. This will prevent other lines close to the area from being rubbed out too.

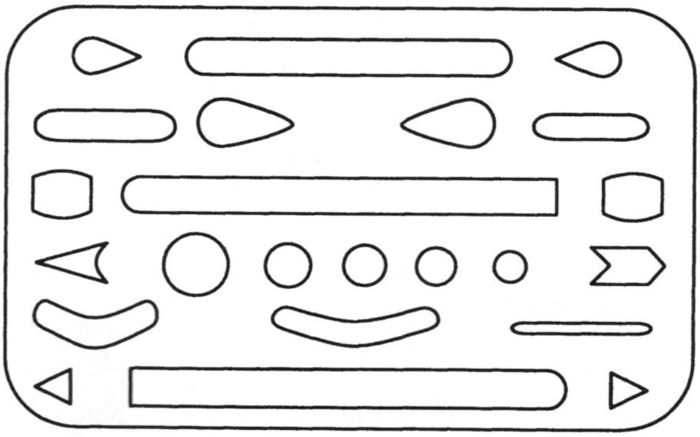

Fig 1.14

2.9 Adjustable compass or large bow and its application

The length of the legs of the compass you use should be approximately 200 mm. Qualified draughtspersons use a **small spring bow** for drawing very small circles, but you do not need one for the purposes of this course.

The compass or large bow is used for drawing **circles and arcs**.

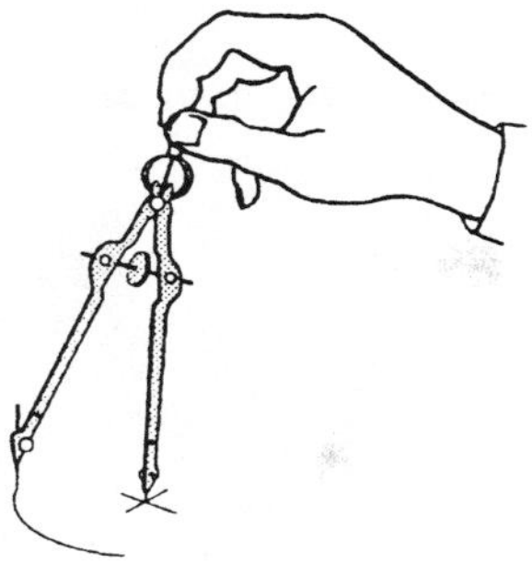

Fig 1.15 Drawing a circle or arc

To draw arcs, you should use the type of point illustrated in fig 1.16 because it gives more accurate results and is easier to maintain than most other types. The needle point should have the shouldered end out and should be adjusted 10 mm beyond the end of the split sleeve.

First sharpen the outside of the lead on a fine file or sandpaper pad to a long flat **bevel** which is approximately 6 mm long. Finish the sharpening process with a rotating motion to reduce the width of the point.

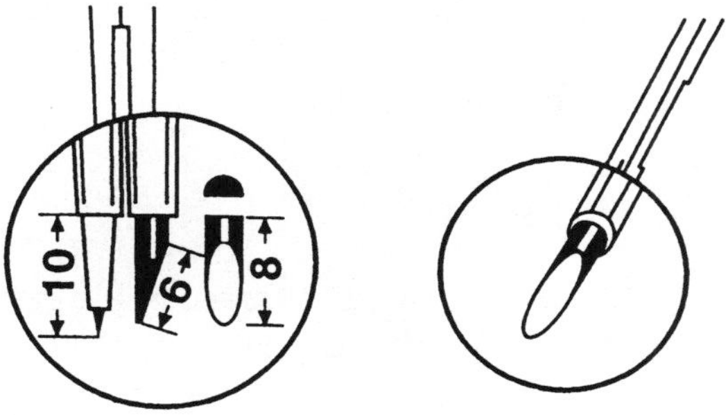

Fig 1.16 Compass point

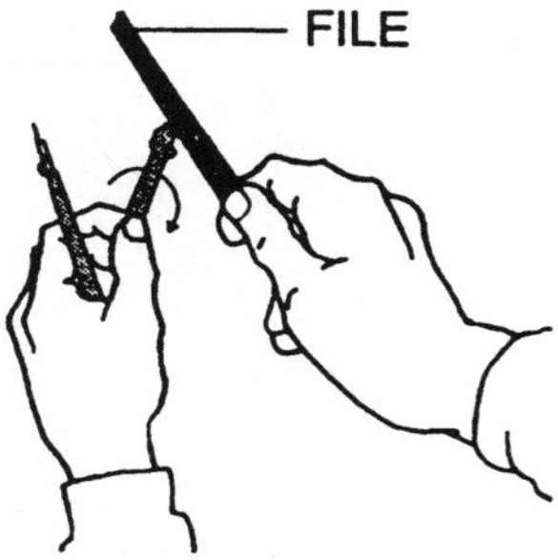

Fig 1.17 Sharpening your compass lead with a file

2.10 Sandpaper pad or file

The sandpaper pad or file are used to sharpen compass and pencil leads.

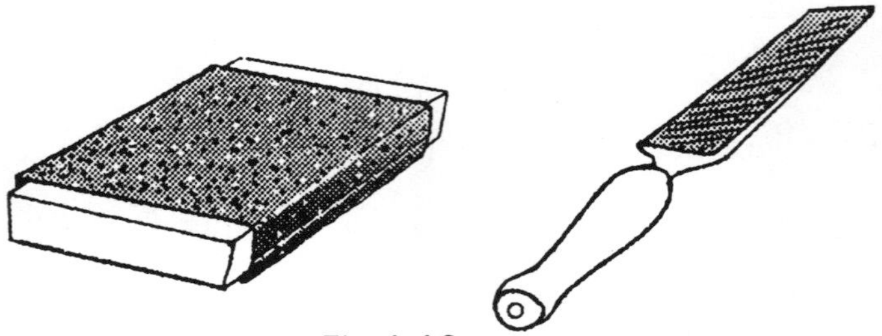

Fig 1.18

2.11 Dust cloth or brush

When working with drawings, cleanliness is essential. Use a dust cloth or brush to help keep your instruments and drawings clean.

2.12 Other instruments

There are other instruments such as protractors (180° and 360°), radius stencils and dividers which are not mentioned here. The reason for this is that only qualified draughtspersons use them for advanced drawings.

2.13 Scale rulers (metric)

The most commonly used scale ruler is the triangular ruler because it has the largest number of scale ratios.

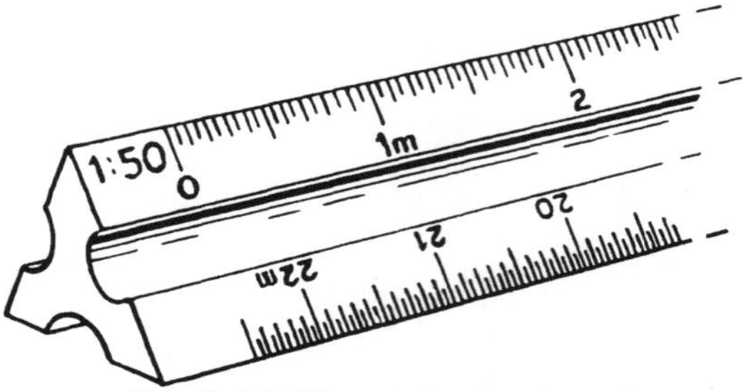

Fig 1.19 Triangular scale rule

Full scale — *Reduced scales* — *Enlarged scales*

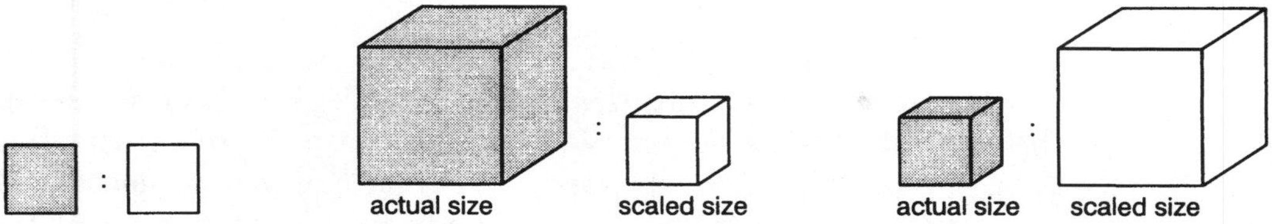

Full scale	*Reduced scales*	*Enlarged scales*
1 : 1	*1 : 2 (twice as small)*	*2 : 1 (twice the size)*
	1 : 3	*3 : 1*
	1 : 5	*5 : 1*
	1 : 10 (ten times smaller)	*10 : 1 (ten times the size)*

The scales 1 : 2,5 and 2,5 : 1 may be used when available or where they are suitable.

The second block is the scaled drawing. To help you remember the order, think of reduced scales as $\frac{1}{2}$ and enlarged scales as $\frac{2}{1}$. The first number is the size of the drawing relative to the object, and is always placed on top of the equation. It is called the numerator. If the scale is 5 : 1, then you know $\frac{5}{1}$ means the drawing will be 5 times the size of the object.

Activity 5

Study the picture of each instrument and make sure that you know the name and function of each item. For example, no. 9 is a dust cloth and it is used to keep the drawing clean. Practise this with all 12 terms. Draw up a list of terms next to the numbers in your roughwork book. Compare your answers with those in the self-evaluation guide.

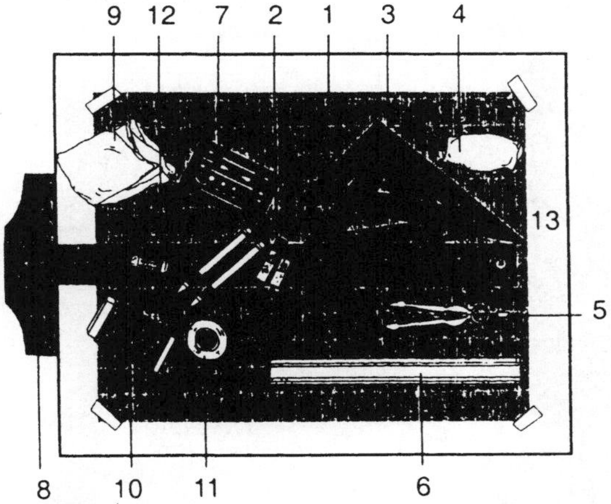

Fig 1.20 Identify the instruments

3 Essential do's and don'ts of neatness in drawing

Neatness is a personal habit. Grubbiness on drawings is usually due to graphite from the pencil. It is the result of smearing the graphite on the lines already drawn. Keep the drawing clean from the beginning. Keep your hands and equipment clean, and graphite- and dust-free. Some precautions are illustrated in fig 1.27. Neatness is not accidental.

Wash hands frequently – using soap

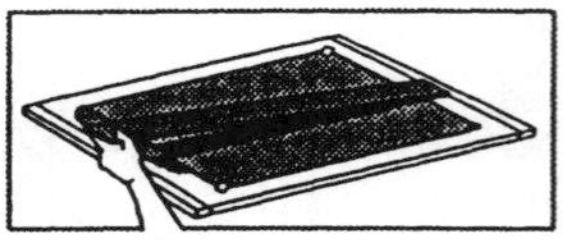

To move T-square, tilt blade up

To move triangle, pick up with fingernail

Never sharpen pencil over drawing

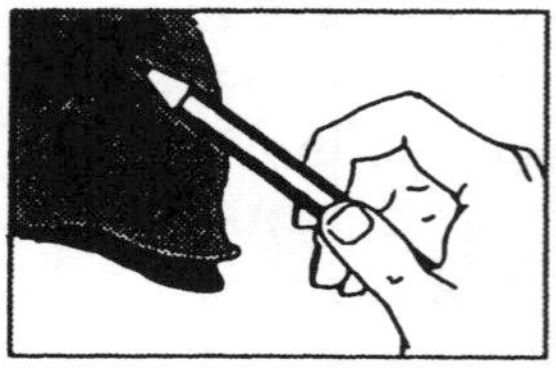

Remove graphite from pencil with clean cloth

Keep sandpaper pad or file in envelope

Never place articles on drawing

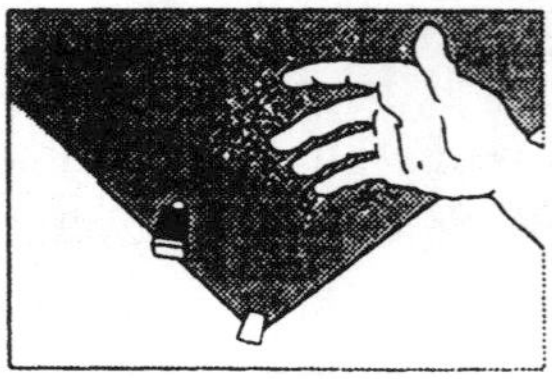

Do not rub off eraser crumbs with hand . . .

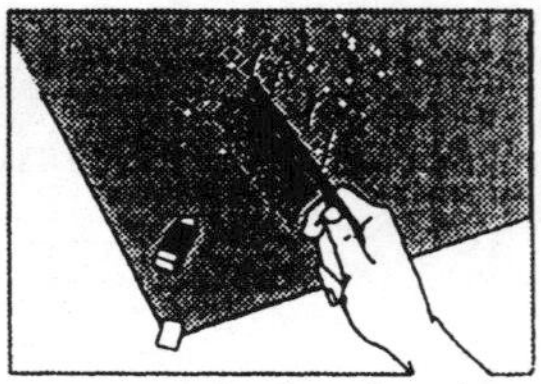

. . . use a dusting brush or flick with clean cloth

Avoid perspiring on drawing

Do not rub fingers through hair

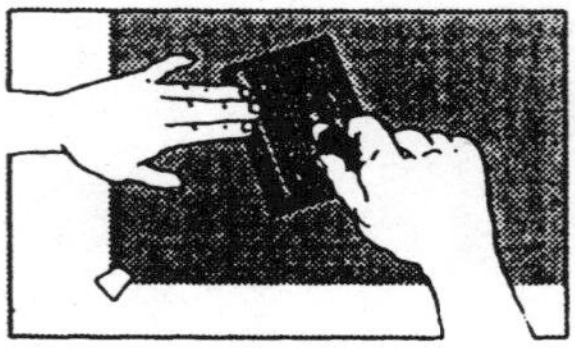

When erasing, protect other lines with erasing shield

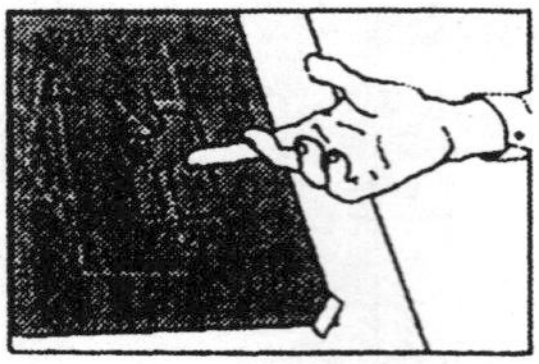

When discussing – touch drawing only with back of fingernail

When lettering, place paper under hand

Always keep T-square and triangles clean

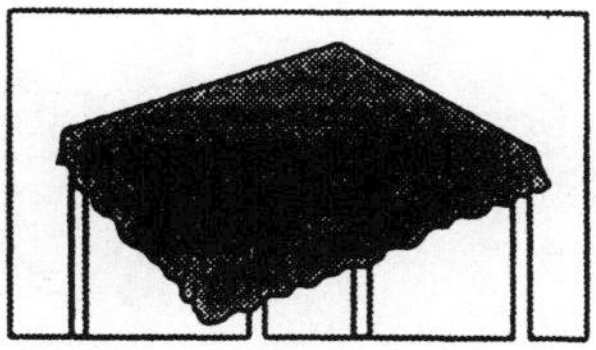

When not drawing, cover board with cloth . . .

. . . or with sheet of paper

Rolling a drawing smears the lines. Carry flat in a large envelope.

Fig 1.21

4 Summary

In this first unit you were introduced to the different basic types of drawing instruments you will be using for this course. You must be able to identify and use these drawing tools correctly, neatly and accurately, since you will need to use these techniques and skills as you work progressively through each unit. In the next unit we will discuss the 'alphabet of lines', which will provide you with further basic insight into drawing.

Self-evaluation

1 Identify each piece of equipment by filling in the correct name next to the number. You should be able to name **all** the instruments correctly. If you find that you cannot recall all of them correctly, practise the activity at the end of section 2.13 until you can.

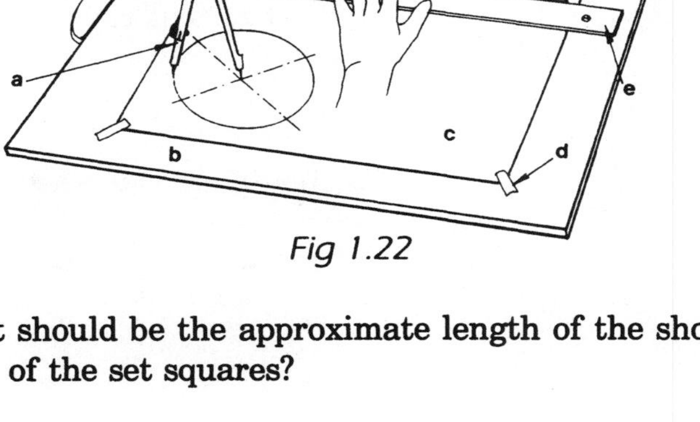

Fig 1.22

2 What should be the approximate length of the shortest sides of the set squares?

3 Why is it advisable for beginners to speak to experienced draughtspersons?

4 What kind of pencil point would you use for drawing straight lines?

5 What is another name for the rounded, pointed shape of a pencil lead?

6 Can you think of reasons why clutch pencils are more practical than ordinary pencils.

7 Explain in your own words why you should not roll up a drawing, but should rather carry it flat in a large envelope. Try to think of at least four reasons.

8 Practise using your different instruments so that you become familiar with how to use them. Begin to apply the habits required to keep your work neat.

Answers

Activity 1

1 perpendicular
2 vertical
3 horizontal, parallel
4 inclined

Activity 2

Practise and make sure that you can draw what is required. Consult with your lecturer, tutor or fellow students to make sure your techniques are correct.

Activity 3

1 conical

2 A conical point is round and by rotating the pencil in your fingers you actually **keep the point sharp** and pointed so that you can draw a thin line.

Activity 4

It prevents smudging; the eraser stays cleaner; the paper is less likely to tear. These are all possibilities and as you become more experienced you will be able to think of more.

Activity 5

This activity is designed to help you improve your ability to **recall** information. Once you have memorised the instruments and their names, cover up the names and test whether you are able to name them correctly. You can test yourself in this way with all your work.

1 Clutch pencil
2 Lead containers
3 Set squares
4 Cleaning pad
5 Compass
6 Scale ruler
7 Erasing shield
8 T-square
9 Dust cloth
10 Sandpaper pad
11 Masking tape
12 Eraser
13 Drawing board

Self-evaluation

1 **a** Compass
b Drawing board
c Paper (A3)
d Masking tape
e T-square
f Set square
g pencil

2 200 mm

3 This would save them from buying expensive instruments which are unnecessary unless the learner intends to become a professional draughtsperson. Experienced draughtpersons can also help evaluate drawings and give advice to beginners.

4 wedge point

5 conical, symmetrical point

6 **a** Saves the learner from having to sharpen pencils, which keeps work neater.
b No need to keep buying ordinary pencils.
c Do not have to carry a large number of pencils around.

7 **a** Lines can be smeared.
b Drawings could be crushed.
c You can carry a larger number of sheets.

d The sheets are better protected against damage.

e It is easier to select and take out the required sheet.

f Sheets are difficult to look at because they keep curling up.

8 It is very important that you **practise** using your drawing instruments correctly and neatly.

Unit 2

Application of the alphabet of lines

Study objectives

After studying this unit, you should be able to draw and identify the following lines:

◇ Outlines, also called object lines

◇ Centre lines

◇ Dimension lines

◇ Extension lines

◇ Leader lines

◇ Construction lines

◇ Projection lines

◇ Guide lines

◇ Dashed lines to show hidden detail

◇ Hatched lines or section lines

◇ Other lines not commonly used

- cutting plane lines or sectional plane lines
- break lines
- phantom lines

1 Introduction

In the previous unit you learnt about the drawing equipment you will be using in this course. In this unit we are going to look at lines, which are an important part of drawing.

Drawing is a **graphic language** used by engineers and draughtspersons. It is made up of symbols, dimensions, notes and different types of lines which transmit meaning.

Written language has letters of the alphabet to form words and sentences. The language of drawing uses a number of different types of lines, called the **alphabet of lines**, to transmit the meaning of a drawing.

It is important that you familiarise yourself with the language of drawing. This language follows the standards set by the South African Bureau of Standards Code of Practice for Engineering Drawing Practice (metric units) (SABS 0111).

When you execute an engineering drawing, you should always keep the following in mind:

- ◊ Correctness
- ◊ Neatness
- ◊ Accuracy
- ◊ Speed
- ◊ Correct layout of the drawings on a drawing sheet.

The quality of lines should meet the requirements of the SABS code.

All lines on a drawing should be of **consistent density** and **reflectance**. This means that the lines must be drawn straight and true and should not vary in width/thickness or appear to be darker in certain places. To achieve this, the pressure of the pencil on the paper must be the same all along the line.

Line types

The linework for this course follows the SABS 0111 Code of Practice for Engineering Drawing and its alphabet of lines, which is listed in table 2.1.

Table 2.1

LINE	*DESCRIPTION*	*GENERAL APPLICATIONS*
A	Continuous thick (straight or curved)	A1 Visible outlines A2 Visible edges
B	Continuous thin (straight or curved)	B1 Imaginary lines of intersection B2 Dimension lines B3 Projection lines (extensions) B4 Leader lines B5 Hatching B6 Outlines of revolved sections in place B7 Short centre lines B8 Bending lines

LINE	*DESCRIPTION*	*GENERAL APPLICATIONS*
BB	Continuous thin and feint	BB1 Construction lines BB2 Guidelines
C	Continuous thin freehand	C1 Limits of partial or interrupted views and sections, if the limit is not a chain
D	Continuous thin (straight) with zigzags	D1 Break line
E	Dashed thick	E1 Hidden outlines E2 Hidden edges
F	Dashed thin	F1 Hidden outlines F2 Hidden edges
G	Chain thin	G1 Centre lines G2 Lines of symmetry G3 Trajectories
H	Chain thin, thick at ends and changes of direction	H1 Cutting planes
J	Chain thick	J1 Indication of lines or surface to which a special requirement applies
K	Chain thin double-dashed Phantom lines	K1 Outlines of adjacent parts K2 Alternative and extreme positions of movable parts K3 Centroidal lines K4 Initial outlines prior to forming K5 Parts situated in front of the cutting plane

Activity 1

Look at the line types in table 2.1. Answer the following questions:

1 How many alphabetical line types are there?

2 What is the thickness of line A?

3 What is the thickness of the other thinner lines?

4 How many applications of the lines are there?

2 Application of the alphabet of lines

In this unit we will briefly discuss the lines listed in the study objectives. Each line will be illustrated and will appear in order according to the frequency with which we use it in this course.

For **object lines** or **outlines**, use a 0,5 mm clutch pencil with a 2H lead; for **all other lines**, a 0,3 mm clutch pencil with a 2H lead will be suitable. This is to ensure that the thicker and darker outlines can always be distinguished from all the other lines in terms of the code of practice for drawing.

2.1 Outlines or object lines (line type A)

The shape of an object is described by thick (heavy) lines known as visible outlines or object lines. These lines are always drawn thick (heavy) and solid so that the outline or shape of the object is clearly emphasised on the drawing.

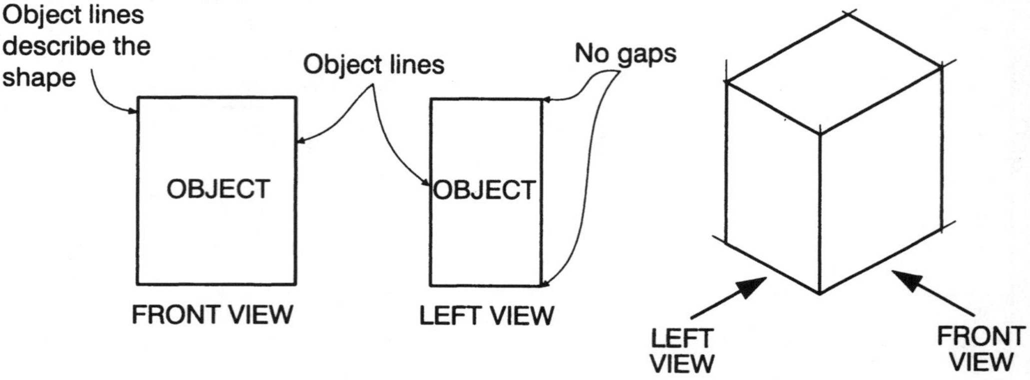

Fig 2.1

These outlines are the most important lines on a drawing.

- ◇ Joining these outlines is very important. There must be no gaps where they meet.
- ◇ They should not overlap.
- ◇ When joining a straight line to an arc, the lines must flow into each other and be of the same contrast.

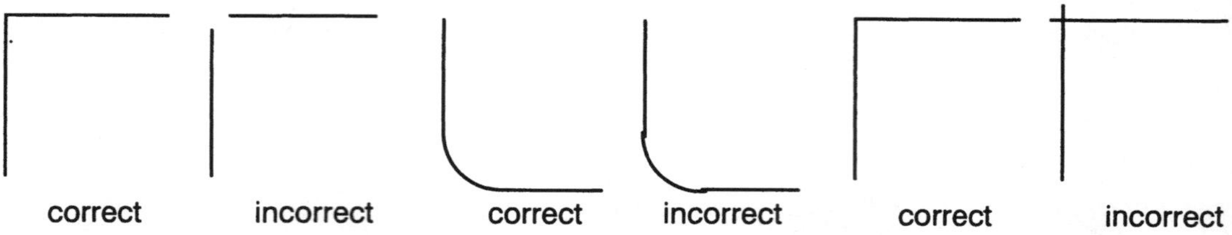

Fig 2.2

2.2 Centre lines (line type G1)

A centre line is drawn as a thin (light), broken line of alternating long and short dashes. Centre lines are used to indicate the centre of a whole circle or a part of a circle, and also to show that an object is symmetrical about a line.

The symbol ℄ is used for a centre line and is derived from the capital letters C (centre) and L (lines). To clarify working drawings, centre lines can be indicated with this symbol.

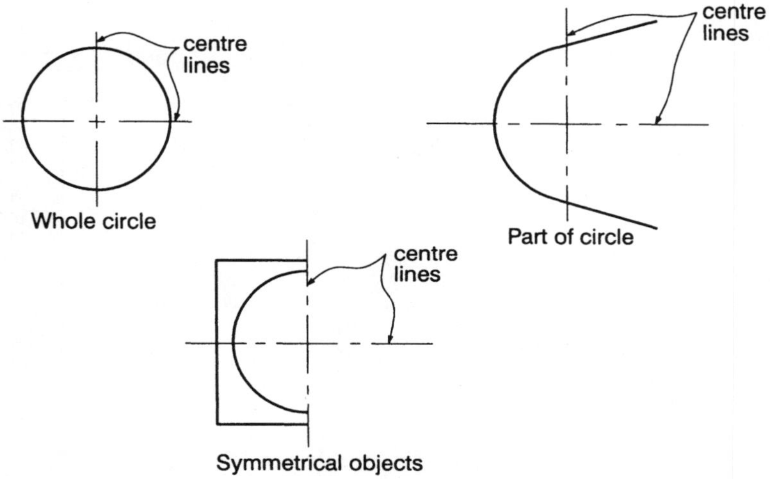

Fig 2.3

Remember the following about centre lines:

◇ Pay special attention to the correct use of these lines. The proportions are as follows:
  - long line 15–20 mm
  - dash line about 2 mm
  - space in between approx. 1 mm

  Keep in mind that these are **approximate lengths** and need not be measured.

Fig 2.4

◇ Centre lines are usually the **first** lines drawn when commencing a drawing.

◇ The centre line should extend slightly beyond the outlines of the drawing.

◇ Do not extend centre lines to adjacent views, i.e. views next to each other.

2.3 Dimension lines (line type B2)

Dimension lines are thin lines ending with arrowheads. The tips or points of these arrowheads indicate the exact distance referred to by a dimension placed on top of the line. (More information on arrows will be given in the next unit.)

The dimension itself must not touch the dimension line. There must be a gap of ±1 mm.

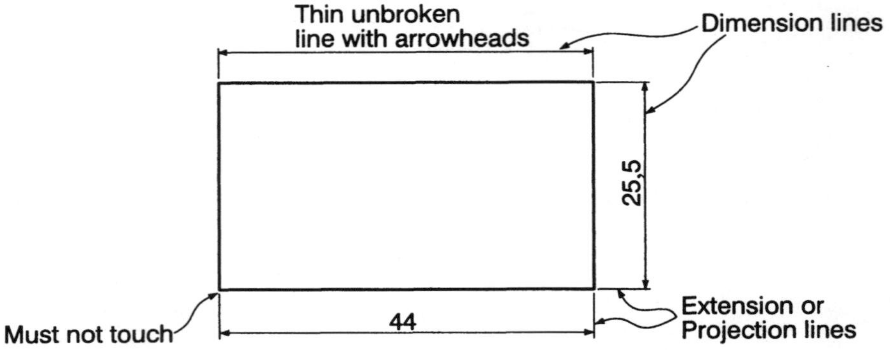

Fig 2.5

2.4 Extension lines for dimensions (line type B3)

Extension lines are used in dimensioning to show the size of the object. Extension lines are thin solid lines which extend away from an object at the exact places between which dimensions are to be placed.

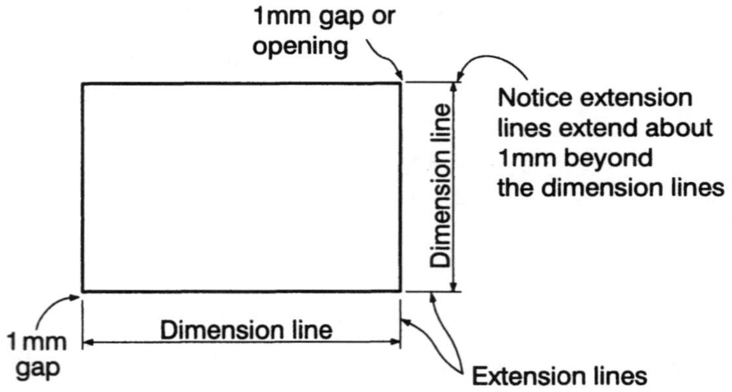

Fig 2.6

The extension line may not touch the object. There must be a gap of approximately 1 mm.

2.5 Leader lines (line type B4)

For a leader line to indicate where a dimension or note is to be inserted, a continuous thin line should be used, terminating in an arrowhead or dot.

An arrowhead should terminate or end on a line; where space is limited, a dot should be located within the outline of the component.

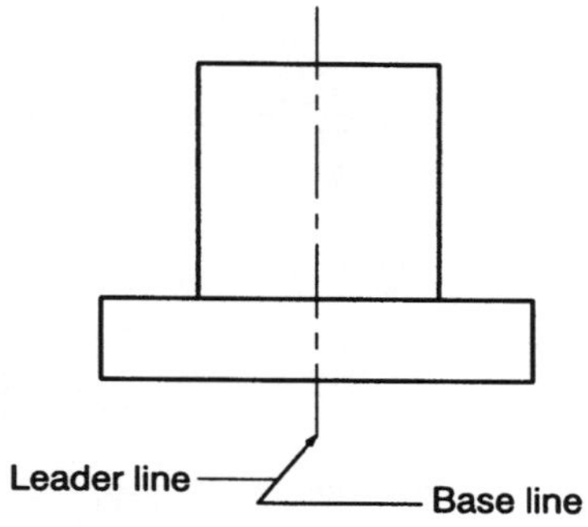

Fig 2.7a Leader lines terminating in an arrowhead

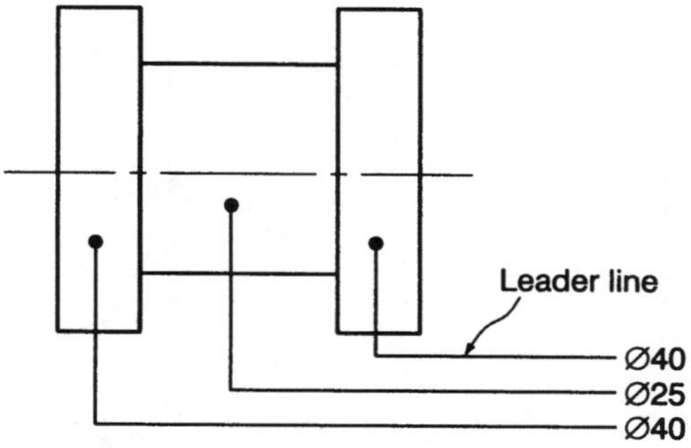

Fig 2.7b Leader lines terminating in a dot (limited space)

Leaders (leader lines) will be discussed in more detail in unit 3.

2.6 Construction lines (line type BB1)

These lines are used to build up a drawing and they are continuous, thin and very faint lines. They also used for geometrical constructions.

Construction lines, if drawn correctly, need not be erased. **They should not be visible from more than half a metre away**. Construction lines will be discussed in more detail later on in the course (see unit 5).

2.7 Projection lines (line type B)

These lines are used by draughtspersons to establish the relationship of lines and surfaces in one view with corresponding points in other views. Projection lines are thin, unbroken lines projected from a point in one view to locate the same point in another view. The mitre indicates where you change direction when projecting. Projection lines do not appear on finished drawings, except where a part is complicated and you have to show how certain details on a drawing are obtained.

When you have finished with the projections, the projection lines may be erased.

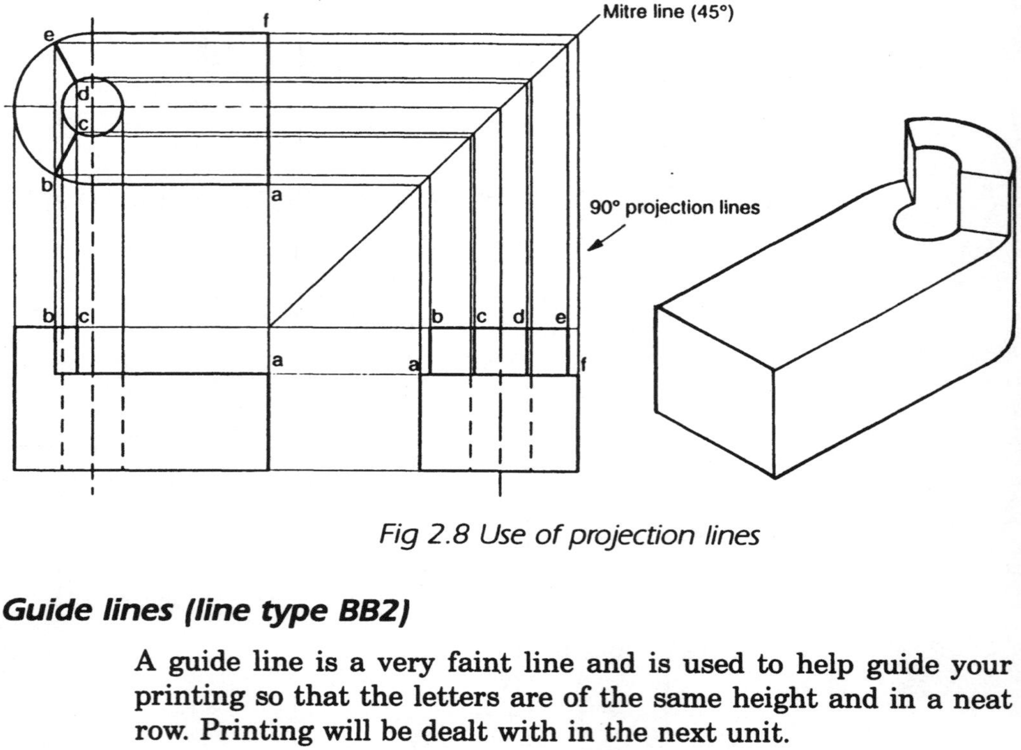

Fig 2.8 Use of projection lines

2.8 Guide lines (line type BB2)

A guide line is a very faint line and is used to help guide your printing so that the letters are of the same height and in a neat row. Printing will be dealt with in the next unit.

2.9 Dashed lines to show hidden detail (Hidden detail lines) (line types E and F)

To be complete, a drawing must include lines which represent all the edges and intersections of surfaces in an object. Many of these lines are invisible to the observer because they are covered by other portions of the object.

This type of line represents invisible edges and surfaces.

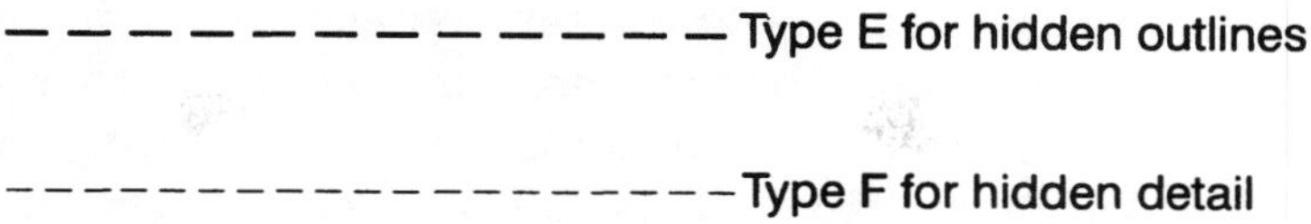

Fig 2.9 Hidden line types

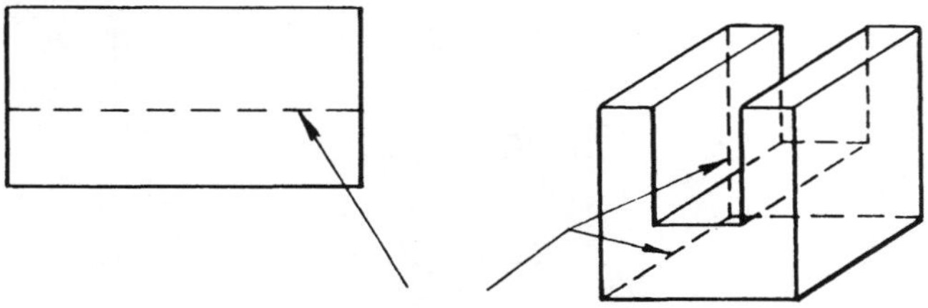

Hidden details are shown in dashed lines

Fig 2.10 Use of hidden detail lines

2.10 Hatched lines or section lines (line type B5)

These lines are continuous and thin, preferably drawn at an angle of 45° to the axis or main line. The spacing between the hatched lines may vary according to the size of the section, but all the lines within a section must be equally spaced.

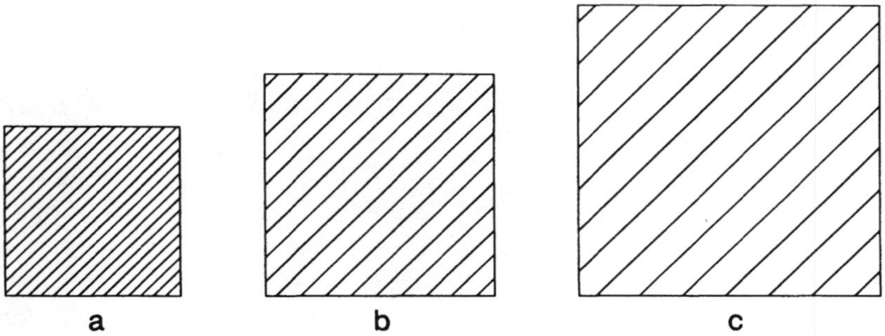

Small sections 1–2 mm
Medium sections 3 mm
Large sections 5–7 mm

Fig 2.11

The use of these lines will be discussed in more detail in unit 7.

2.11 Other lines

In addition to the ten common types of lines used in this course, the alphabet of lines also includes other types such as **cutting plane lines** or **sectional plane lines, break lines** and **phantom lines** to indicate adjacent parts and alternative positions and lines for repeated detail. These less frequently used lines are **found in more advanced drawings** and will be described in greater detail when they are used in later drawings.

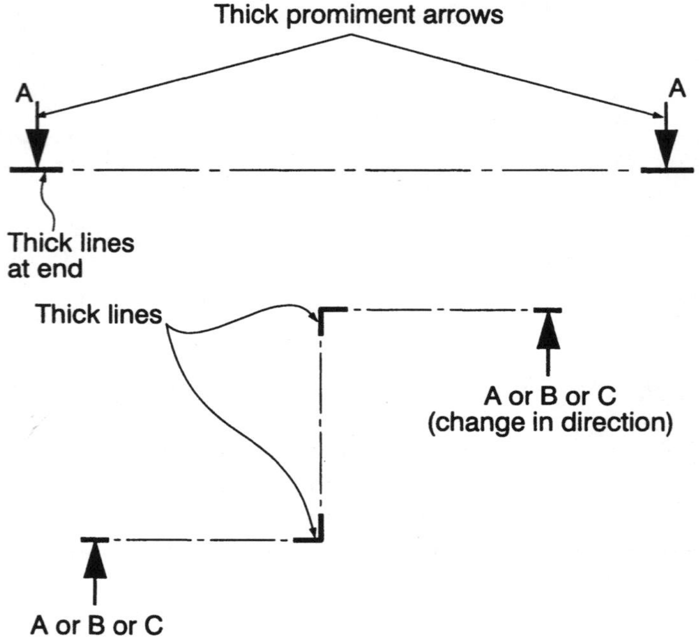

Fig 2.12a Cutting planes or sectional planes (Line type H)

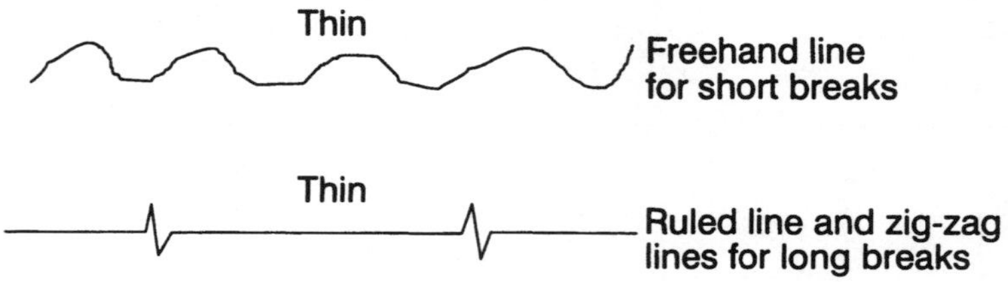

Fig 2.12b Break lines (Line types C and D)

Fig 2.12c Phantom lines (Line type K)

Phantom lines are broken lines of long dashes followed by two short dashes and indicate adjacent parts and alternative positions.

Activity 2

Complete the following table which summarises the 14 lines discussed in this course. Six lines have been drawn and summarised to serve as examples. See if you can complete the table.

Table 2.2

LINE TYPE		EXAMPLES	REMARKS
1 Outlines	A1	1	Thick, heavy visible outlines
2 Centre lines	G1		
3 Dimension lines			
4 Extension lines	B	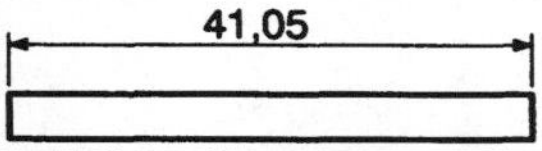 4	Used in dimensioning to show the size of an object. Thin and solid – extend away from object – must not touch object.
5 Leader lines			
6 Construction lines			
7 Projection lines	B	7	Continuous, thin – projected from point in one view to locate some point in another view. May be erased once finished, unless complicated.
8 Guide lines			
9 Dashed lines			
10 Hatched lines			
11 Cutting plane lines	HI	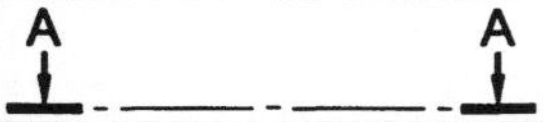 11	Thick, prominent arrows – thick lines at end.
12 Break lines (short)	C		
13 Break lines (long)	D		
14 Phantom lines		14	Thin – double-dashed – indicates adjacent parts and alternative positions.

3 Summary

Having completed unit 2, you will realise how important it is to be able to identify and apply the alphabet of lines correctly and accurately. In the next unit we will be revising the third category of basic drawing skills, namely lettering, figuring and dimensioning.

Self-evaluation

Study the following drawing and answer the questions on the lines illustrated below.

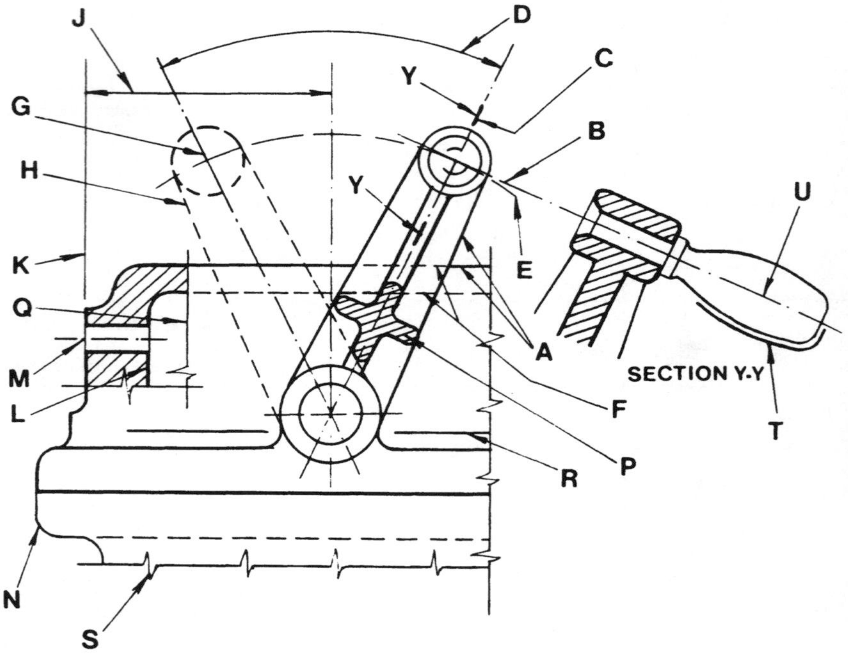

Fig 2.13

Questions

1 What kind of line is A?

2 What kind of line is B?

3 What kind of line is D?

4 What kind of line is F?

5 What kind of line is G?

6 What kind of line is C?

7 What kind of line is Q?

Answers

Activity 1

1 11

2 0,5 mm

3 0,3 mm

4 28

Activity 2

If you complete this table correctly, you can use it as a recall pattern or memory map for revision purposes. Write down only the relevant information.

Self-evaluation

1 An outline

2 A centre line

3 A dimension line

4 A hidden outline

5 A centre line

6 A cutting plane or sectional plane line

7 A break line (ruled line for long breaks).

You can use any drawing in a textbook and practise naming and identifying the different lines. At first you will always have to refer to the SABS 0111 code, but with practice you will be able to draw and identify the lines with ease. The trick, however, is to practise these lines and drawings often. An hour's practice a day, five days a week for six months is an example of the time you will need to become skilled if you are doing this course through distance education. This practice is in addition to the time you spend studying the contents of this course and doing your assignments.

Unit 3

Lettering, figuring and dimensioning

Study objectives

After studying this unit, you should be able to do the following correctly and accurately in terms of the requirements of the drawing code of practice:

- ◇ Prepare the drawing sheet
- ◇ Apply the lettering and figuring
- ◇ Position the title and scale in relation to each other
- ◇ Position the title and scale on a drawing sheet
- ◇ Insert the dimensions on a drawing
- ◇ Insert all the different types of dimensions on a working drawing
- ◇ Insert tolerance dimensions
- ◇ Insert dimensions on keyways
- ◇ Identify and interpret the different symbols for orthographic projections
- ◇ Use leaders
- ◇ Insert item numbers (also called part numbers)
- ◇ Insert a parts list or item list
- ◇ Insert machining symbols.

1 Introduction

Drawing, as you learnt in the previous unit, is actually an engineering language consisting of an 'alphabet of lines' instead of letters. Apart from a graphic representation of the shape of a part, machine or structure, there is more to drawing than simply drawing lines. It is also very important to remember that drawing requires that you **print** the descriptions, figured dimensions, notes on material and its finishing, as well as the drawing's descriptive title and scale. All this additional work must be done in a style that is **legible**, **uniform** and **neat** in appearance.

In order to finish a drawing, it is extremely important that you apply the basic lettering, figuring and dimensioning principles of drawing, as required.

Poor lettering and figuring can spoil a good drawing. In this unit we are going to show you the correct applications of lettering, figuring and dimensioning.

2 Lettering, figuring and dimensioning in drawing

2.1 Preparation of the drawing sheet

Draw a borderline 10 mm from the edge of the drawing sheet all around the **sheet** and a **name-block** located at the **bottom right-hand corner** (see fig 3.1).

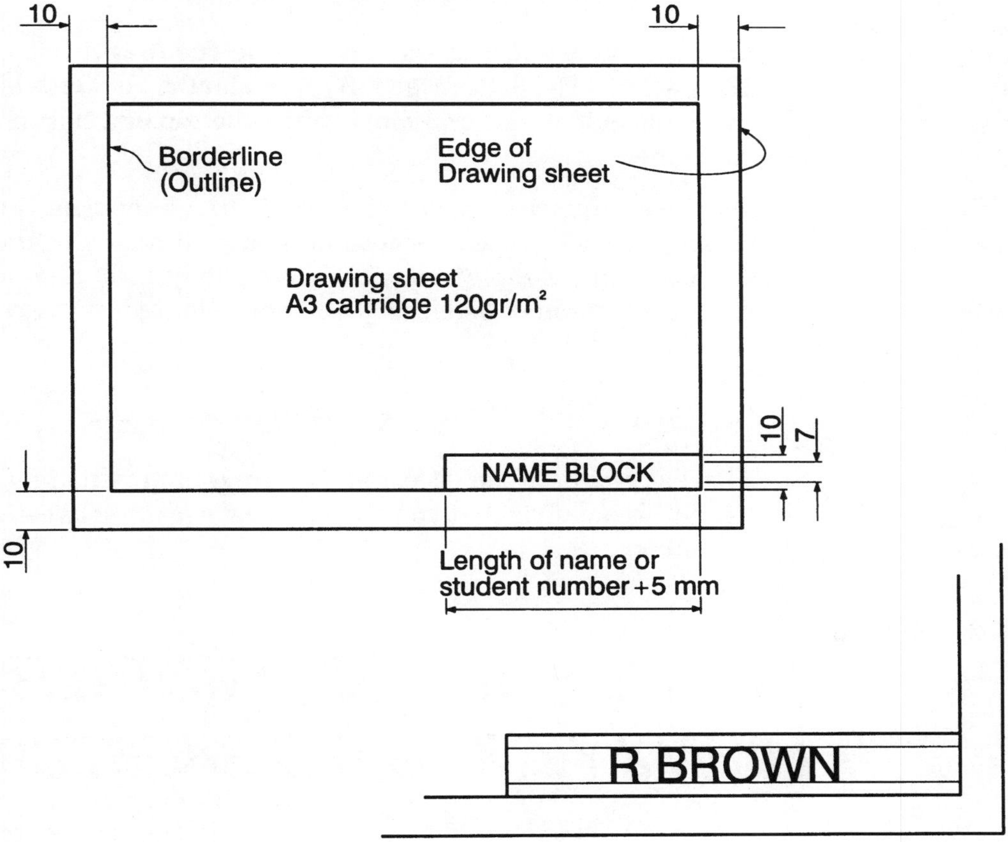

Fig 3.1 Preparation of the drawing sheet

The dimensions shown in fig 3.1 are very important and must **always** be used in every drawing.

Type of paper used = A3 cartridge paper 120 g/m^2

Name block: This is where you print your **surname and initials or student number**. The lettering must be **7 mm high**. Make use of the correct types of letter and figures, as discussed in the next section.

2.2 Lettering and figuring

The true shape of a part or mechanism may be described accurately on a drawing by using combinations of lines and views. Added to these are letters and figures, which supply additional information, and the dimensions.

As we have pointed out before, poor lettering and figuring can spoil an otherwise good drawing and you should therefore get as much practice as possible in these skills.

Lettering and figuring should be done **freehand** with an **H or 2H pencil**. The letters and figures should be written between two feint guidelines, and must touch the top and bottom of these guidelines.

An almost square style of letter and figure, known as **Gothic lettering**, is widely used because it is easy to read and the individual letters are simple enough to be made quickly and accurately. Gothic letters may be either **inclined** (slanted) or **vertical** (upright).

Inclined (slanted) letters and figures

The distance between letters and figure is approximately $\frac{1}{3}$ of their height. Remember that you should always print between two faint guidelines.

Fig 3.2a

Upright (vertical) letters and figures

The distance between letters and figures is approximately $\frac{1}{3}$ of their height. Again, print between two faint guidelines.

Fig 3.2b

Heights of letters and figures used in engineering drawing

Title (name of object) = 7 mm high

Scale used = 5 mm high

Dimensions and notes = 3 mm high
(see fig 3.3)

2.3 Positions of the title and scale relative to each other

NOTE: THE SCALE IS CENTERED UNDER THE TITLE

Fig 3.3

The **title** is the **name** of the object drawn.

Scale is the type of scale used, e.g. full scale or scale 1 : 1, which means the drawing is to be drawn the same size as the object.

Dimensions and notes will be dealt with later on in this unit.

2.4 Positioning of the title and scale on the drawing sheet

Because we have to draw more than one drawing on a drawing sheet for economic reasons, the title and scale must be placed as follows below each layout drawing of an object.

You must first draw the drawing and then position the **title** and the **scale** as shown.

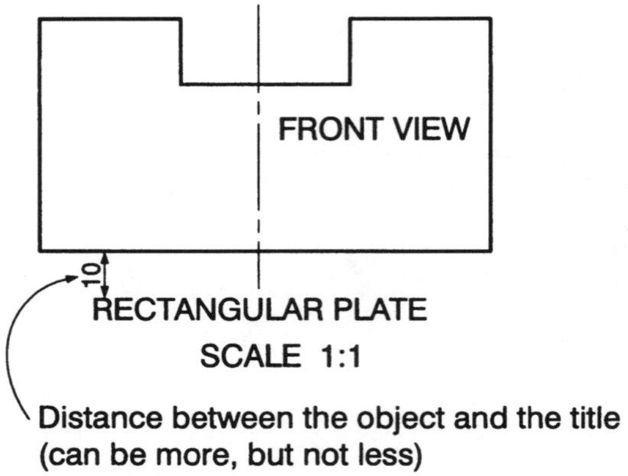

Fig 3.4a For a single view of an object

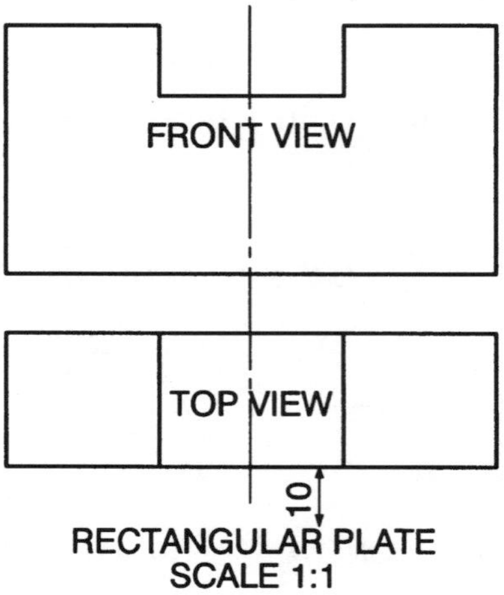

Fig 3.4b For two views of an object

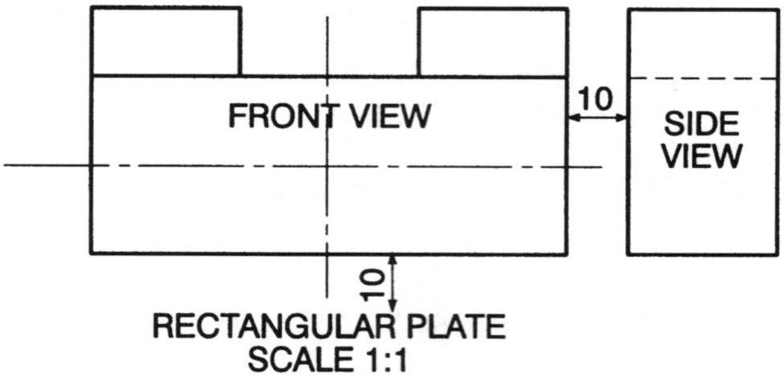

Fig 3.4c Alternative for two views

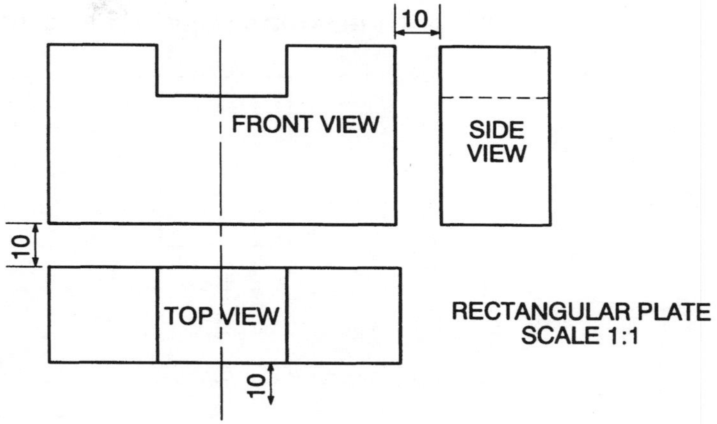

Fig 3.4d For three views of an object

Remember, the information inside the drawings consists of **notes** so the height of the letters must be **3 mm**.

2.5 Dimensioning on a working drawing

A drawing consists of several types of lines which are used singly or in combination to describe the shape and internal construction of an object or mechanism. However, to guide the construction or machining of a part, the drawing must include dimensions which indicate the exact size and location of surfaces, indentations, holes and other details.

The lines and dimensions, in turn, are supplemented by **notes** which give additional information; for this reason the **Code of Practice for Engineering Drawing SABS 0111 – Part 1–1990** is very important. In this unit every dimension will have to comply with this **code**.

All **dimensions** are given in **millimetres** on a drawing, unless otherwise specified.

Inserting dimensions (vertical and horizontal)

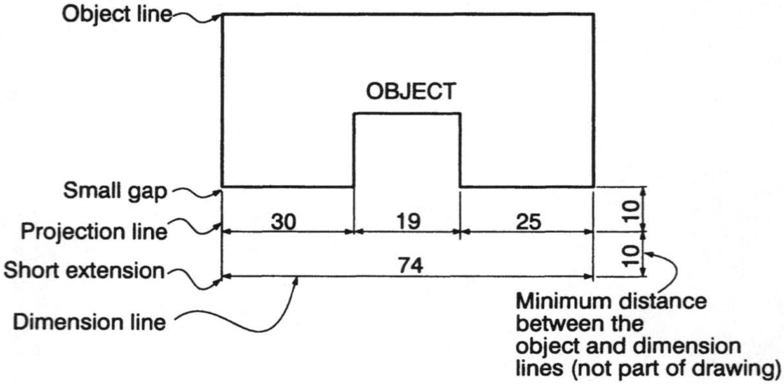

Fig 3.5a Dimensions below the view of an object

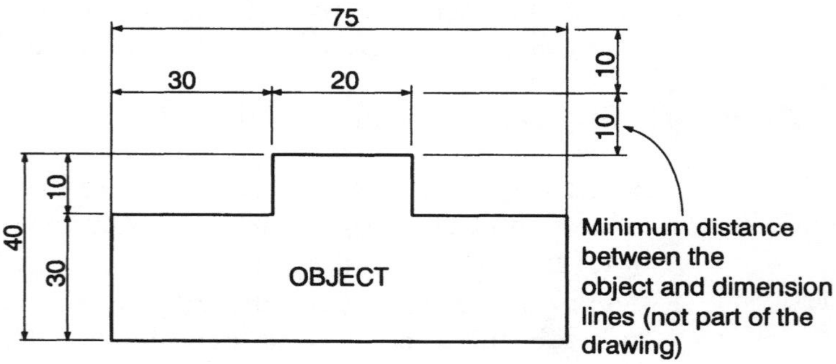

Fig 3.5b Dimensions above the view of an object

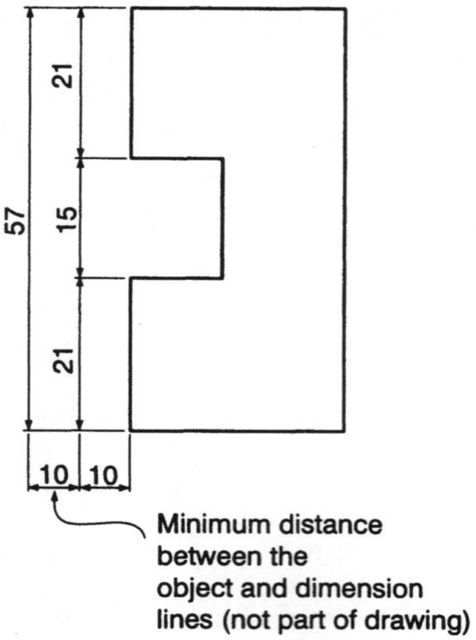

Fig 3.5c Dimensions to the left of the view of an object

Dimensions are oriented in such a way that they are placed on top of the dimension line as read from either the bottom or right hand side of the page.

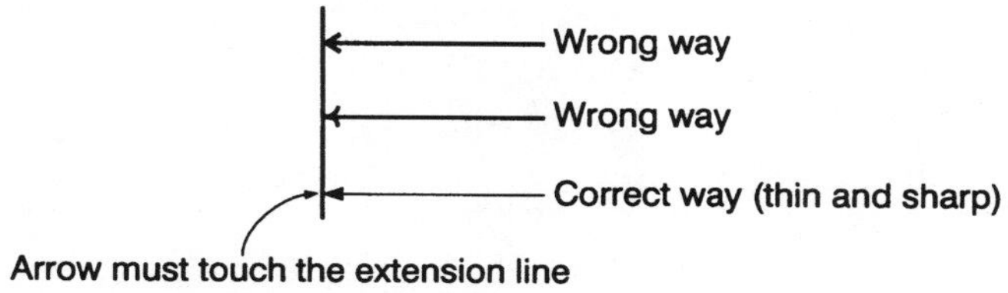

Fig 3.5d The correct and incorrect ways of inserting arrowheads

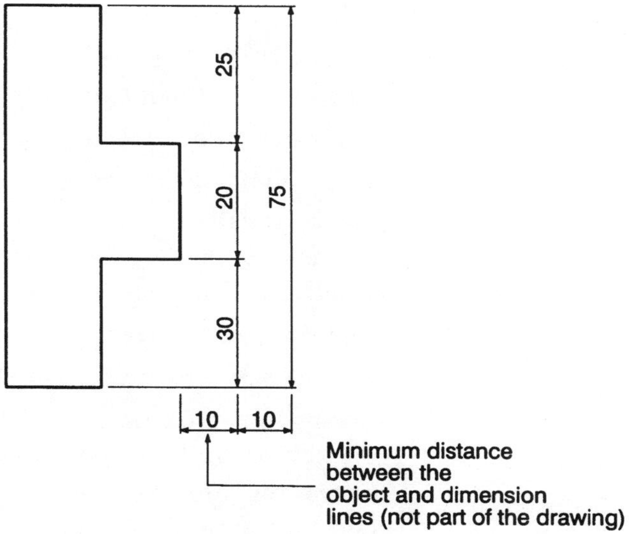

Fig 3.5e Dimensions to the right of the view of an object

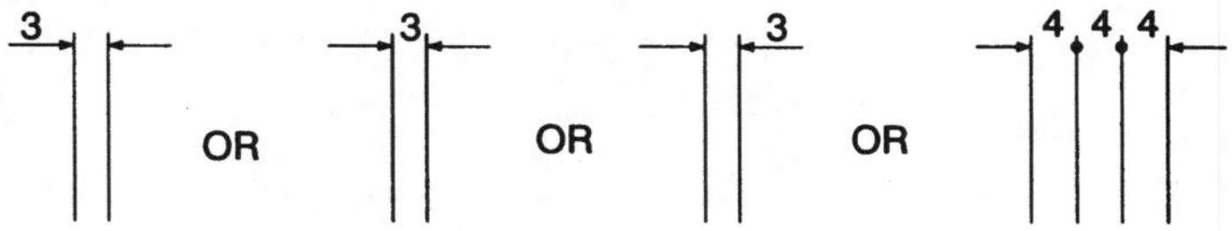

Fig 3.5f Small dimensions

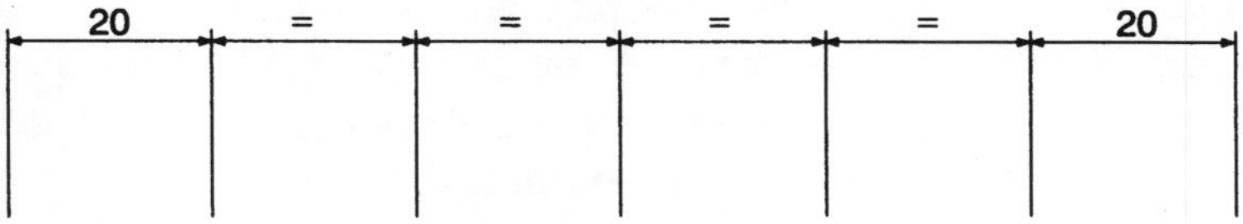

Fig 3.5g Repetition of the same dimensions

This is where a series of the same dimensions must be shown.

Hints on dimensioning (summary)

1. Use a 0,5 mm clutch pencil with 2H lead. (Remember that lettering and figuring are done with either a 2H or HB lead.)
2. Numerals must be 3 mm high.
3. Dimensions not closer than 1 mm to the dimension line.
4. Dimensions must be not closer than 10 mm to the drawing.
5. All dimensions are in millimetres, unless otherwise specified.
6. Use guidelines to print the dimensions.
7. Indicate each dimension only once.
8. Do not use a centre line as a dimension line.
9. Place dimensions in the centre above the dimension line.
10. **Where possible**, dimensions should be placed outside the drawing.
11. **Distribute** dimensions over different views.
12. Avoid dimensions from dashed lines.
13. Place overall dimensions on the outside of the drawing.
14. Dimensions must be placed so that they may be read from the bottom or right-hand side.
15. Always consider how the part will be manufactured when inserting dimensions.

The points mentioned above are **very important**. You must be able to give a good interpretation of a working drawing using these techniques.

Activity 1

In this activity you will produce review sheets of the material presented so far. These sheets will help reinforce your understanding of the material and will also help you prepare for exams. You will prepare a sheet for each of the sub-sections in section 2. Fig 3.6 shows an example of a review sheet for section 2.1. Prepare similar sheets for sections 2.2 through 2.5:

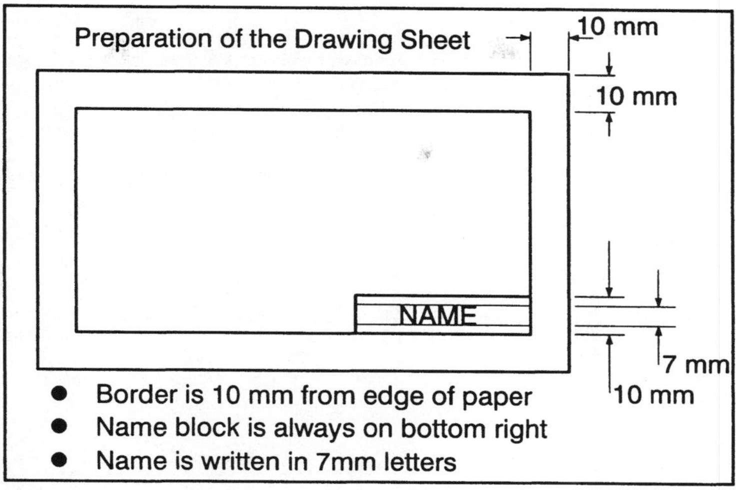

Fig 3.6

2.6 Different types of dimensions on a working drawing

Direction of reading dimensions

Dimensions have to be oriented in such a way that they can be read from the **bottom** or **right-hand** side of the **drawing**.

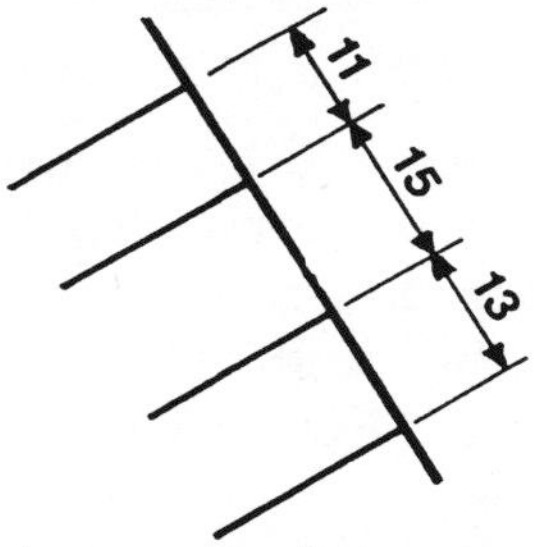

Fig 3.7a Linear dimensions are written parallel to the dimension line

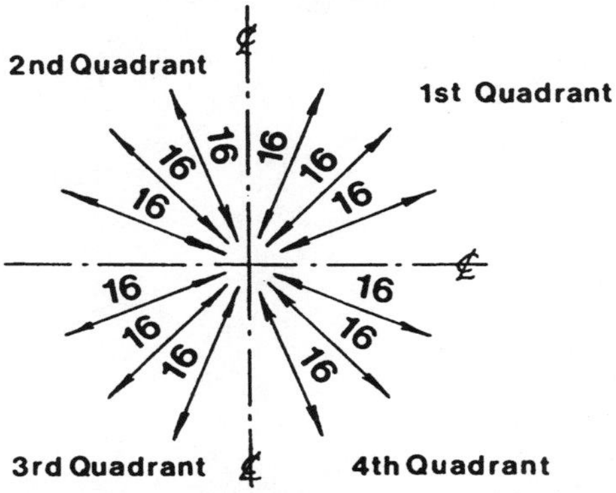

Fig 3.7b Orientation of inclined linear dimensions

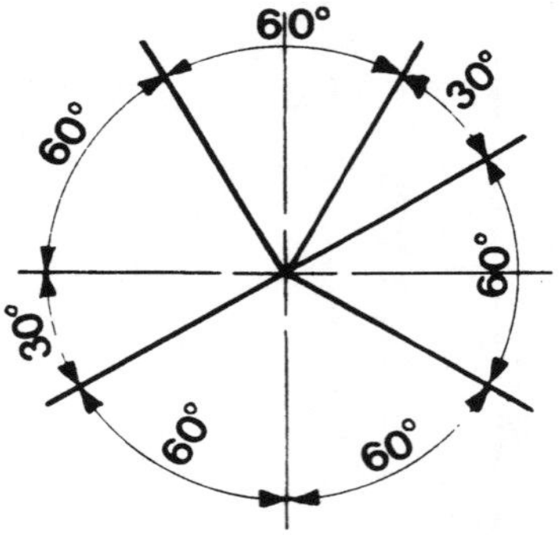

Fig 3.7c Orientation of angular dimensions

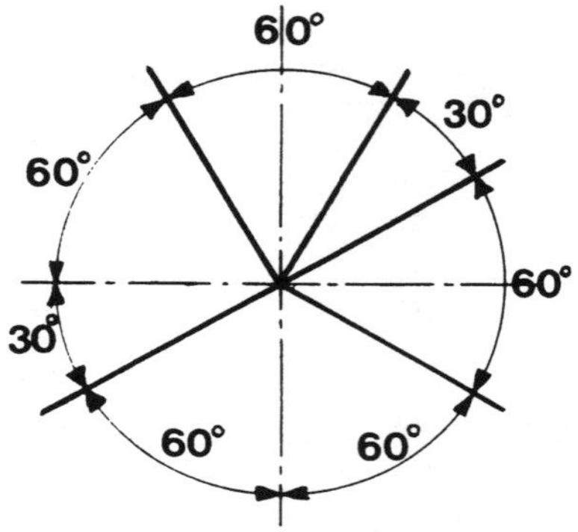

Fig 3.7d Angular dimensions written horizontally

Dimensioning of circles

Circles can be dimensioned by using leader lines to indicate their diameter or radius. Leaders to be in line with centre of circle. Radii are dimensioned almost the same as circles by using only half of the circle. The radius is indicated by an R. The diameter is indicated by the symbol Ø.

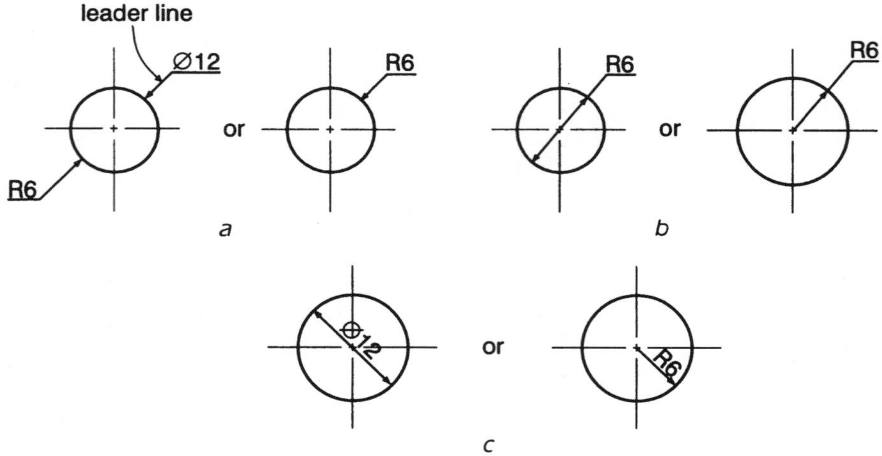

Fig 3.8a–c

Dimensions of side-view diameters on a drawing

Fig 3.9 shows diameters dimensioned in a side view. In this case you read from the **right-hand side** of the drawing and not from the bottom.

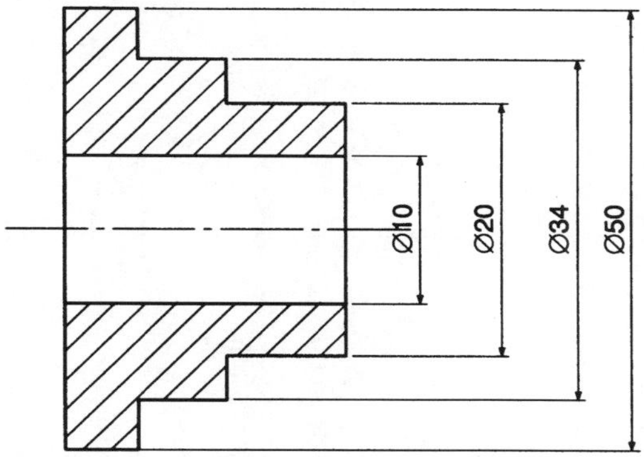

Fig 3.9 Dimensioning of diameters (side view)

It is important to choose the side on which to place the dimension so that it is as close to the diameter as possible, as shown in fig 3.10.

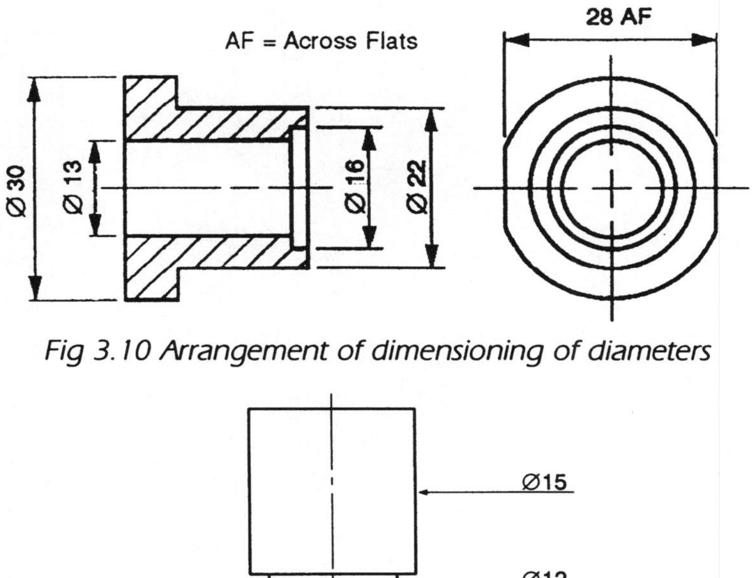

Fig 3.10 Arrangement of dimensioning of diameters

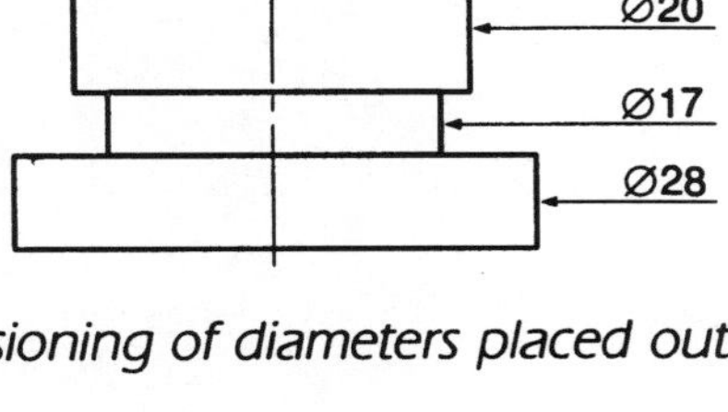

Fig 3.11 Dimensioning of diameters placed outside component

When there is a pattern of holes, it is not necessary to dimension each one. Fig 3.12 shows two methods for dimensioning holes on a pitch circle. The pitch circle diameter is the distance to the centre of the holes.

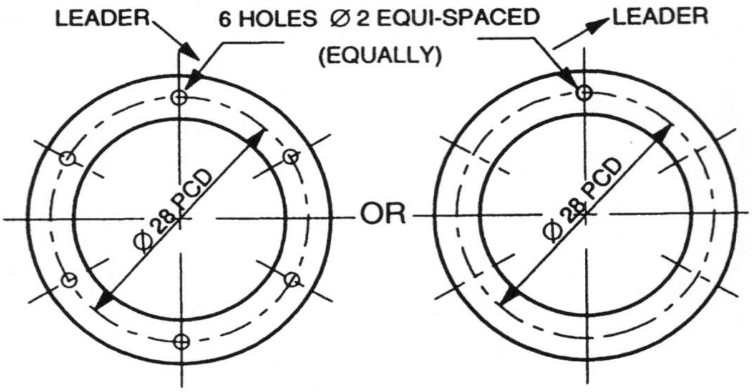

Fig 3.12 Holes on a pitch circle by using leaders

PCD = Pitch Circle Diameter

Dimensioning diameters of spheres

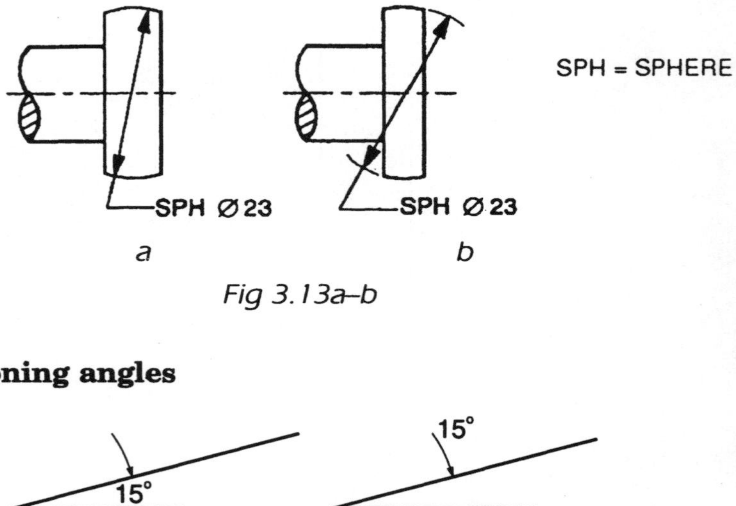

Fig 3.13a–b

Dimensioning angles

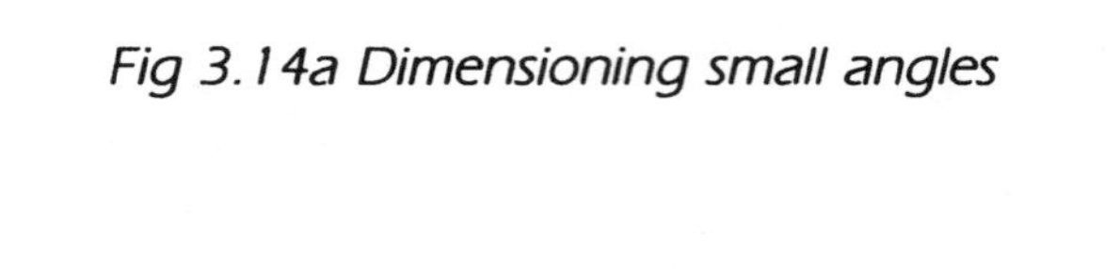

Fig 3.14a Dimensioning small angles

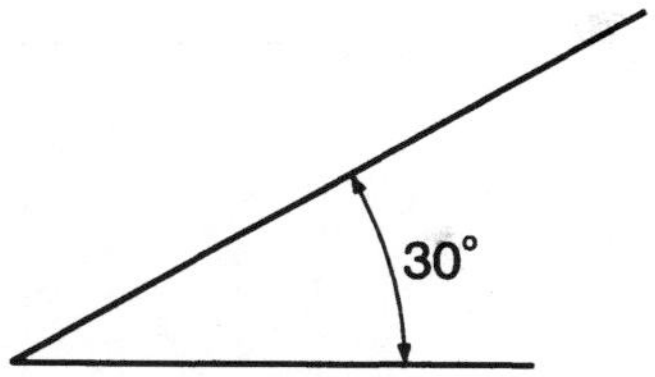

Fig 3.14b Dimensioning large angles

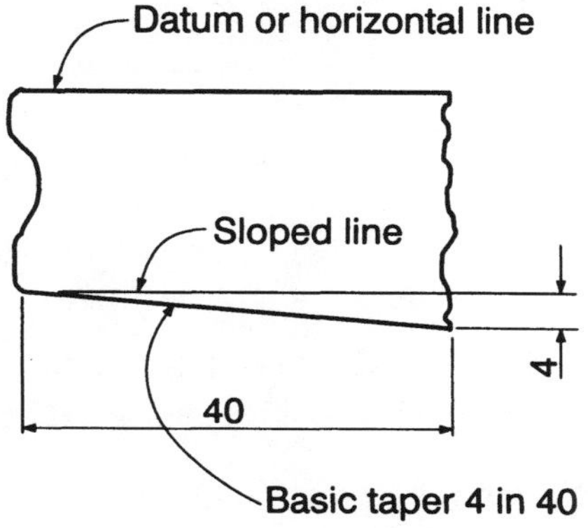

Fig 3.14c Dimensioning a taper (basic)

(This means that for every 10 mm there will be a slope of 1 mm.)

Dimensioning chamfers

Chamfers are used to remove sharp corners. The **edges** slope at 45 ° to the vertical and the horizontal. (In fig 3.15 the **height** of the chamfer is 4 mm.)

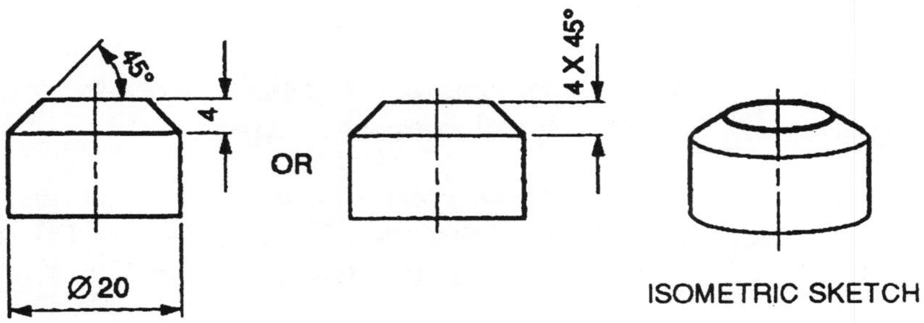

Fig 3.15

Dimensioning countersinks

Countersunk screws or rivets are used where flat surfaces are required. A countersink slopes at 45° to the vertical and the horizontal. (In fig 3.16 the depth of the countersink is 4 mm.)

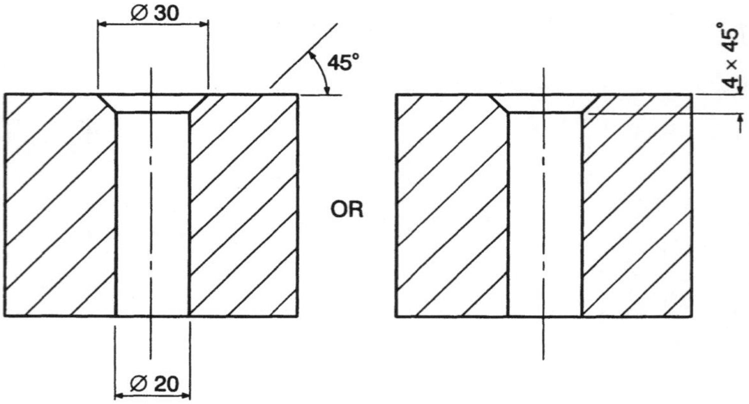

Fig 3.16

Dimensioning regular metric threads (conventional representation)

This matter will be discussed in more detail later on. Here we are merely showing you the correct way to dimension a metric thread.

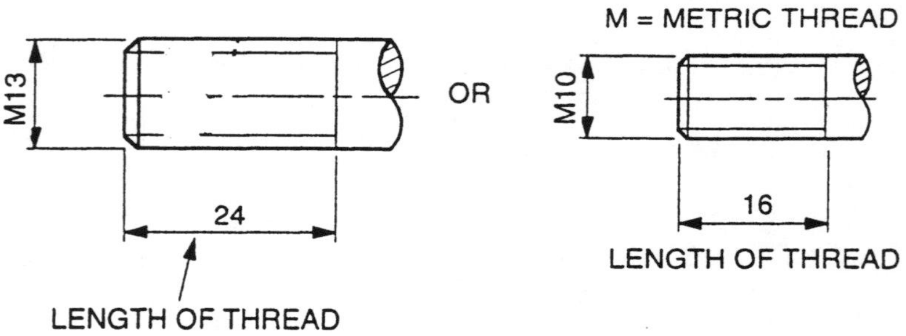

Fig 3.17

Dimensioning special metric threads (conventional representation)

Special threads must indicate the diameter and the pitch (distance between threads). Fig 3.18 shows a thread that is 17 mm in diameter with 2 mm between threads.

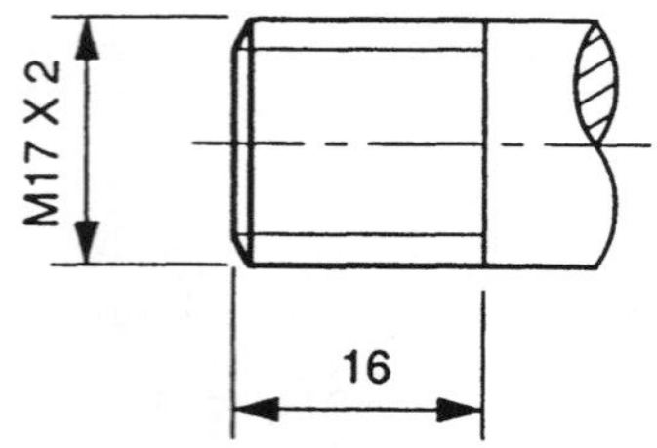

Fig 3.18

Dimensioning a metric threaded hole

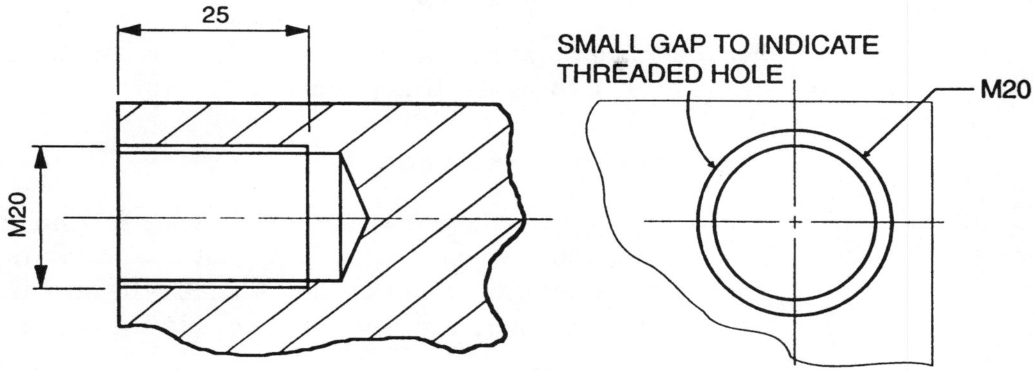

Fig 3.19

Good and bad practice in the use of dimension lines

The introduction of an axis of symmetry, a centre line, can result in a great simplification of dimensions (see fig 3.20).

A centre line or part of an outline should never be used as a dimension line, but can be used as a projection line (see fig 3.20).

This means you cannot use a centre line or dashed line as dimension line. You can use a dimension line only to show a dimension.

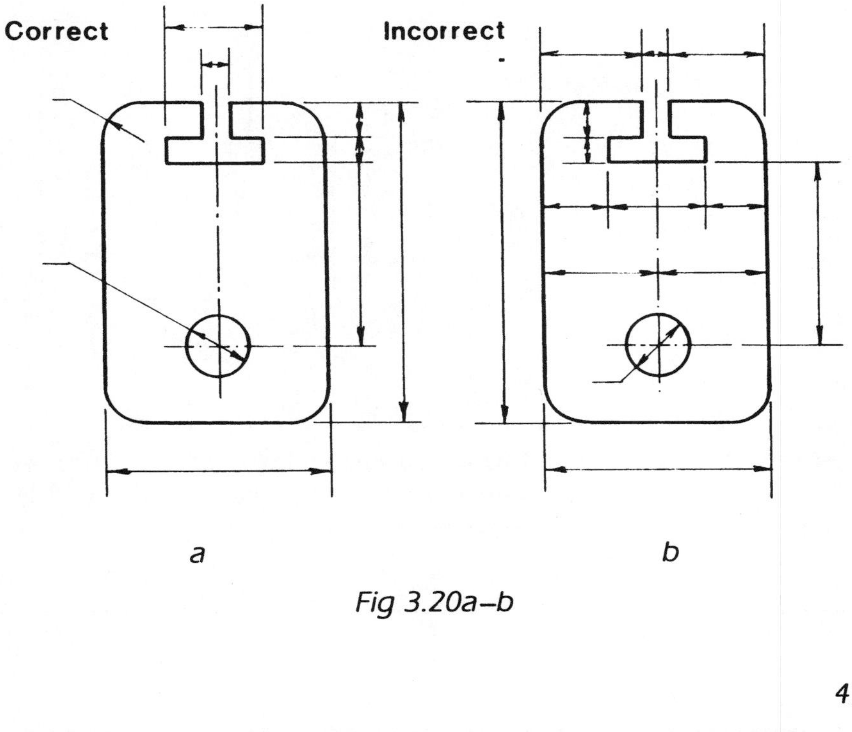

Fig 3.20a–b

Fig 3.20a shows all the dimensions on the outside of the drawing. **Try to use this practice.**

Fig 3.20b shows almost all the dimensions on the inside of the drawing. **Try to avoid this practice.**

Separation of dimensions

The drawing must make it as easy as possible for the artisan to measure the dimensions accurately so that the machining can be done with the minimum of problems. Usually the arcs and circles are milled first, when it is still easy to clamp the object.

It is a good practice to separate dimensions that provide information about the **position** and **size** of a feature (item).

Fig 3.21 gives positional dimensions, marked P, and the size dimensions marked S. This means that some dimensions (P) give the **positions** of the centre lines for the arcs and circles, and others (S) the **size** dimensions for arcs, circles and objects.

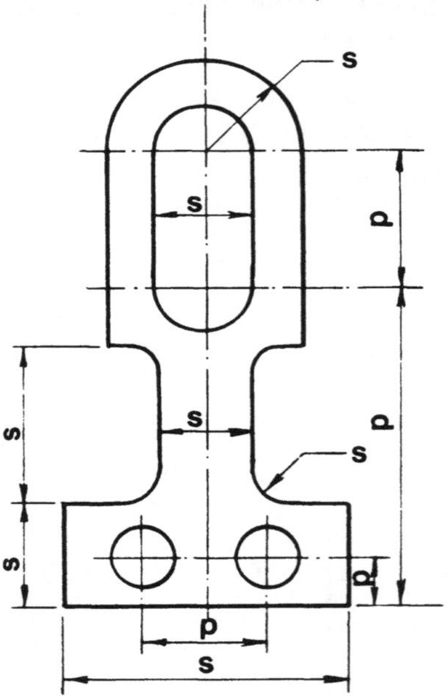

Fig 3.21

Activity 2

Create your own review sheets of the different **types of dimensions**. Refer to activity 1 on how to do this. Place the sheet where you can refer to it often.

2.7 Tolerance dimensions

It is generally impossible to manufacture components to **exact sizes**, therefore tolerances are used to define the maximum limits of size that are acceptable. The magnitude of the tolerance will vary with the class of work.

Tolerancing by specifying both limits of sizes

Tolerances can be indicated by specifying the upper and lower limits of the acceptable size.

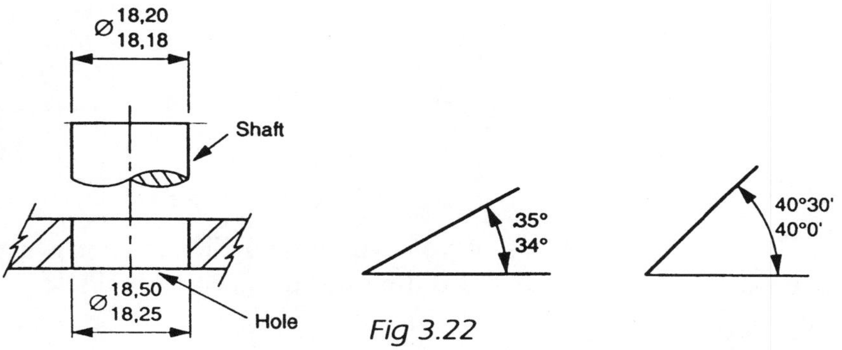

Fig 3.22

Tolerancing by specifying the nominal size and accepted variation

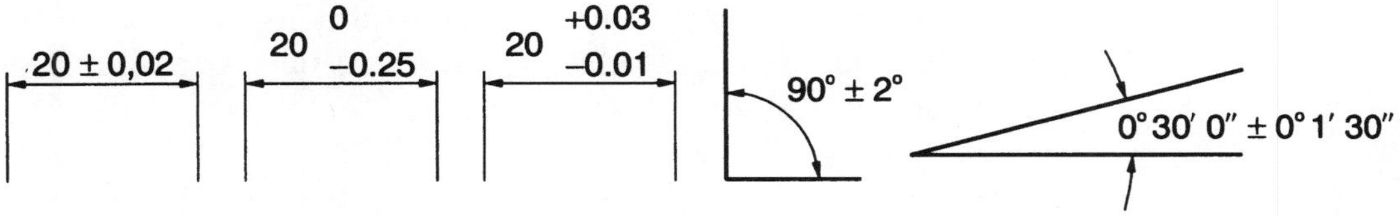

Fig 3.23

2.8 Dimensioning keyways

One of the methods shown below should be used to dimension keyways for square and rectangular keys.

A shaft fits into a hub together with the key which transmits rotary motion. The key prevents the shaft from turning around in the hub. A hub is the central part of the wheel or pulley. Fig 3.24a shows a keyway in a shaft. Fig 3.24b shows a keyway in a hub.

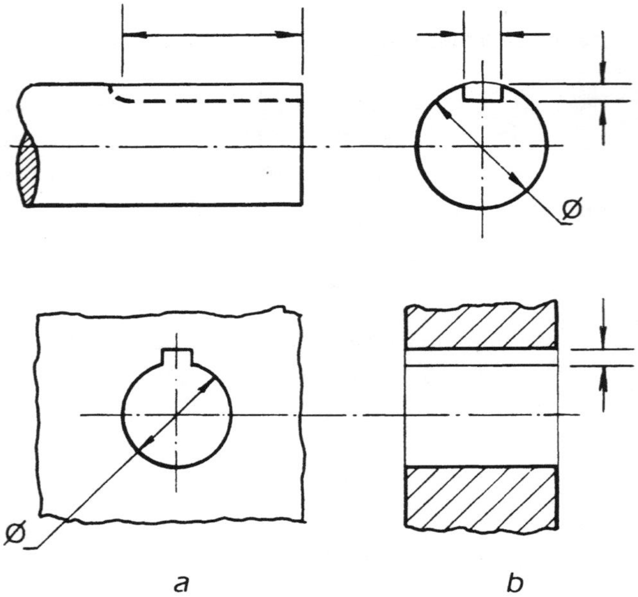

Fig 3.24a–b

For those of you who would like to improve your skills and do more advanced dimensioning, consult SABS 0111 Part 1–1990.

2.9 Leaders

Type of line

A continuous thin line (the same thickness as a dimension line) should be used, terminating in an arrowhead or a dot, for the leader line (leader) to indicate where a dimension or note is to apply. A dot should be located within the outline of the component, where space is limited.

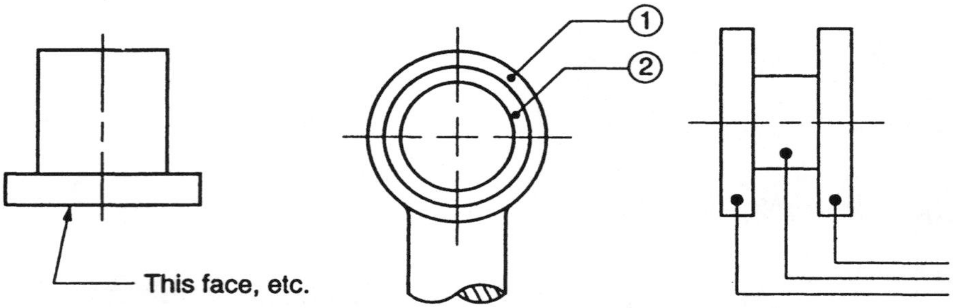

a
Leader terminating in an arrowhead

b
Leaders terminating in dots

c
Leaders where space is limited

Fig 3.25a–c

Leaders that touch outlines

The angle between leader lines and outlines should be as close to 90° as possible.

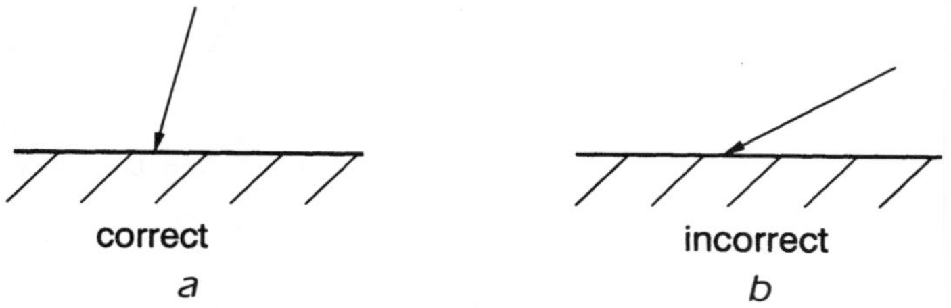

Fig 3.26a–b

Long leaders

Long leaders should be avoided. See the difference between the drawings below. Where all features in a drawing have the same dimension, a single leader can be used. The term TYP indicates that the dimension is typical of all others.

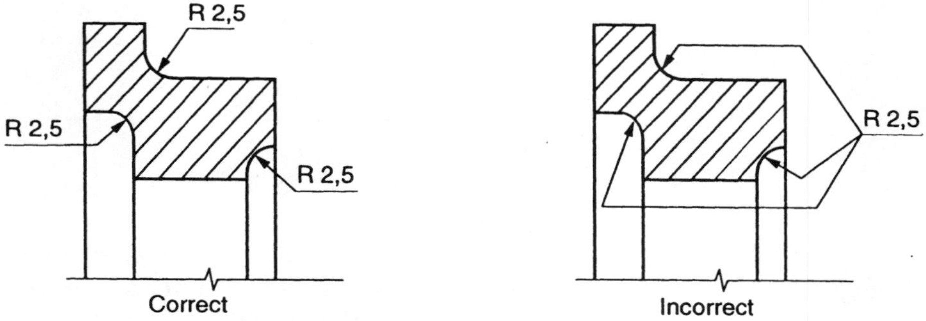

Fig 3.27

2.10 Machining symbols

Machine parts that are **cast** or **forged** normally have rough surfaces. The surfaces that are to be machined should be marked on the drawing by using the appropriate symbol.

The basic symbol is a 60° vee with unequal legs, as shown in fig 3.28a.

If the machining involves the removal of material, then the symbol is as shown in fig 3.28b.

Where machining is not permitted, the symbol is as shown in fig 3.28c.

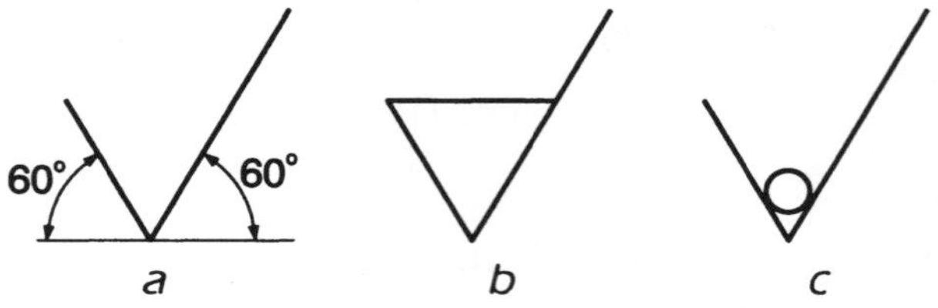

Fig 3.28a–c

The correct way to insert the machining symbols is shown in fig 3.29.

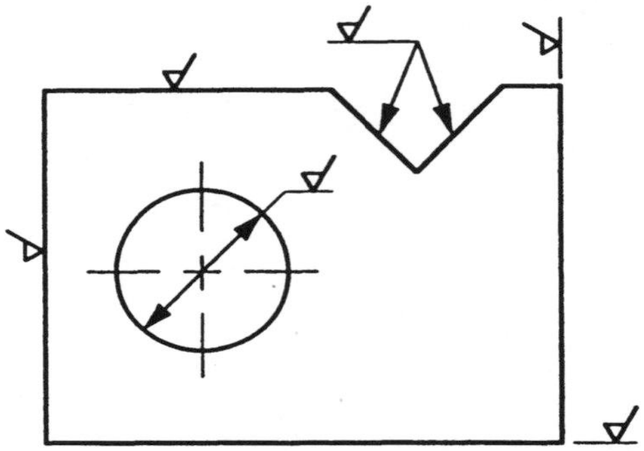

Fig 3.29

Activity 3

As you did in the previous two activities, create your own review sheets – this time, of the work covered in sections 2.7 to 2.13. Make sure you have included all the headings, sub-headings and key facts. Be as creative as you like.

3 Summary

Having studied units 1 and 2, you should now be able to demonstrate how to use the drawing instruments correctly, how and where to **apply** the different types of lines, and how to insert the different types of dimensions.

These first **three units** contain content which explains the **basic rules of engineering drawing**, as set out in the SA Standard Code of Drawing, SABS–0111.

Drawings have to be done according to these accepted standards of practice. Before you can proceed to the next unit, you should be familiar with the contents of these standards.

The only way to acquire the necessary competence in these basic drawing skills is to **practise** continually in your own time, as we have emphasised before.

Self-evaluation

Section 1 (Units 2 and 3)

1 What is an outline? Where is it used?

2 What is an object line?

3 What is a centre line?

4 What is the main function of a centre line?

5 What type of line is used to draw a line of symmetry?

6 Describe a dimension line.

7 Where is an extension line used?

8 Must a dimension line end in a dot?

9 What is a leader line and where is it used?

10 What is a construction line?

11 What type of line is used for hidden detail?

12 What is a cutting plane?

13 Where do you use hatched lines?

14 When you print the title, what letter height should you use?

15 When you print the scale, what letter height should you use?

16 When you print both, what must their positions be relative to each other?

17 Where do you insert the title and the scale for the following?

- **a** one view
- **b** two views
- **c** three views

18 What must the height of a dimension be?

19 How far away must the nearest dimension line be from a drawing?

20 Must the dimension touch the dimension line?

Section 2

The following questions are typical of the kind that you would be asked in an assignment.

Prepare an A3 drawing sheet, as shown under 2.1, and follow the instructions below. Refer to the examples provided in the unit to ensure the correct applications of lettering, figuring and dimensioning.

Question 1

Print the alphabet of letters and numerals shown in section 2.2. You may print them upright or inclined and they must be 7 mm high. The gap between the letter lines must be 5 mm. The forming of the letters and numerals must be **exactly the same** as shown in this unit. (Remember to use your guidelines.)

Question 2

Repeat the exercise in question 1, except that this time the height of the letters and numerals must be 5 mm. The gap between the letter lines must also be 5 mm.

Question 3

Again repeat the exercise in question 1, but now use a height of 3 mm for the letters and numerals. The gap between the letter lines must now be 2 mm.

Question 4

Print the following titles and scales:

Remember to refer to the notes in this unit before you start answering this question.

BEARING
SCALE 1:1

PULLEY
SCALE 1:2

SHAFT COUPLING
SCALE 1:5

PISTON FOR STEAM ENGINE
SCALE: FULL SIZE

BIG END
SCALE 1:1

Question 5

Print the following **notes**: (Remember the height.)

SECTIONAL FRONT ELEVATION
SECTIONAL END ELEVATION
SECTIONAL PLACE
DRILL ALL LOGS 20
CAST-IRON HAND WHEEL
BEARING BUSH
GUN METAL VALVE
BIG END
VALVE SPINDLE
VECTOR DIAGRAM
SPACE DIAGRAM
INCLINED LINE

Answers

Regard activities 1–3 as learning activities which you can use as part of your revision of this subject.

Section 1

1. An outline is used to show the shape of an object clearly and is described as a solid heavy line.
2. An object line is the same as an outline.
3. It is a thin broken line of alternating long and short dashes.
4. Centre lines are used to indicate the centre of a complete circle or arc, and are also used for symmetry.
5. A centre line type.
6. Dimension lines are thin continuous lines ending with arrowheads.
7. Extension lines are used in dimensioning to show.
8. A dimension line ends with an arrowhead.
9. A leader line is a continuous thin line terminating in an arrowhead or dot and is used to indicate where certain dimensions or notes are to be applied.
10. Construction lines are used to build up a drawing and they are continuous, thin and very faint lines.
11. A thin broken line (short dashes).
12. A cutting plane is an imaginary cut through an object.
13. You use hatched lines on a cutting plane.
14. 7 mm.
15. 5 mm.
16. 5 mm gap between them with the scale centred below the title.
17. **a** Symmetrically below the view.

    a Symmetrically below the two views.

    c Below the left or right view.
18. 3 mm.
19. 10 mm.
20. No, it must not touch the dimension line because it may give a false impression of the dimensions.

If, when evaluating yourself using these questions, you do not obtain full marks, we suggest that you go back to your study material and concentrate on those areas of the work with which you are experiencing difficulties. If you are still uncertain after having revised the work, contact either your lecturer, tutor or another student who can help you to understand and master these basic 'tools' of drawing. Remember, we encourage you to consult with other people who can help you improve your comprehension of the subject.

Section 2

Questions 1–7 involve activities that test the basic 'tools' of drawing. These are skills you need to **master**, because if you are not sure about the basics **now**, you will encounter many problems later on.

Unit 4

Freehand sketching

Study objectives

After studying this unit, you should be able to:

- ◇ Explain the difference between **real-life** (field) sketching and **design** sketching
- ◇ Sketch the following without using any instruments:
  - horizontal lines, vertical lines and slanted lines
  - curved lines constructed from **simple geometric forms**
  - arcs connecting straight lines
  - circles
  - irregular shapes
  - isometric views or isometric projections
  - oblique drawings
  - isometric circles on three different planes
  - oblique circles on three different planes
- ◇ Explain the concept of foreshortening and its functions
- ◇ Sketch threaded fasteners and describe their functions
- ◇ Describe and draw locking devices
- ◇ Describe and draw joining methods.

1 Introduction

Many parts or assembled units can be illustrated clearly and adequately by freehand sketches.

Sketching involves drawing without using drafting equipment. It is also a way of conveying ideas rather than a method of making complete drawings. Many people believe that freehand sketching is in fact more important than working drawings because artisans or engineers often have to communicate their ideas through sketches.

The most important aspect of freehand sketching is **good proportion**. No matter how good the line work, if the proportions are not correct, the sketch will be poor.

The material provided in this study unit will help you master the **freehand skills** you will need for sketching.

What is the value of sketching?

- ◇ Sketching helps you to sort out ideas and decide which design looks best.
- ◇ It gives you something to show other people in order to get their advice and benefit from their ideas.
- ◇ It serves as good preparation for technical drawing when you use instruments.
- ◇ It is of great value in preparing and planning the arrangement of drawings on a sheet.

2 Real-life (field) sketching and design sketching

In the engineering field, there are two main classes of sketching: **real-life** (field) sketching and **design** sketching.

Real-life sketching

Real-life sketching takes place when you sketch components, machines and structures, often at remote sites far from the office. You will usually draw these parts in field books or on A4-size clipboards. Remember, rather take too many measurements than too few!

Design sketching

Design sketching is involved in the initial stage of any design. For this you need to have imaginative ideas, resourceful insight and perspective sketching skills.

- ◇ Choose a sharp, **soft pencil**, usually B or 2B, for **design** sketches. Develop a light touch for initial outline.
- ◇ Use **slightly harder** H or HB pencils for real-life or field sketches. A soft eraser and suitable paper are essential.

3 Freehand sketching

3.1 Sketching horizontal lines

To sketch horizontal lines, the pencil should be held lightly on the paper and inclined at an angle. The line should be drawn in one continuous motion.

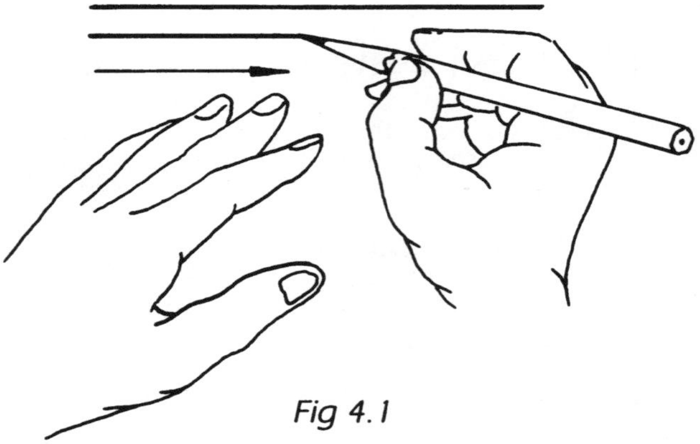

Fig 4.1

3.2 Sketching vertical lines

Vertical lines are drawn by pulling the pencil towards you.

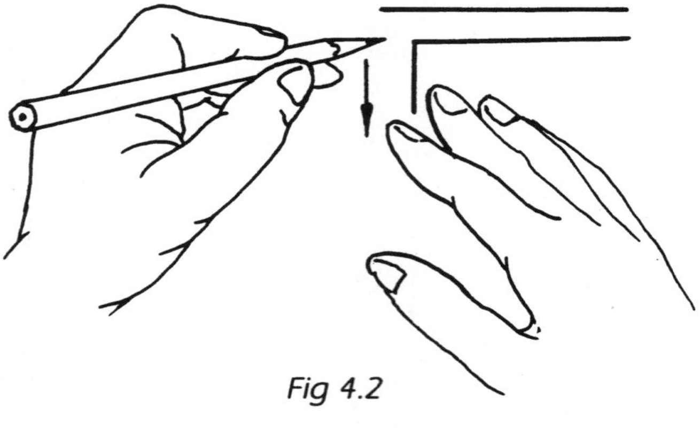

Fig 4.2

3.3 Sketching slanted lines

Slanted lines should be drawn at an angle, using the same method as horizontal lines. It may be helpful to turn the page.

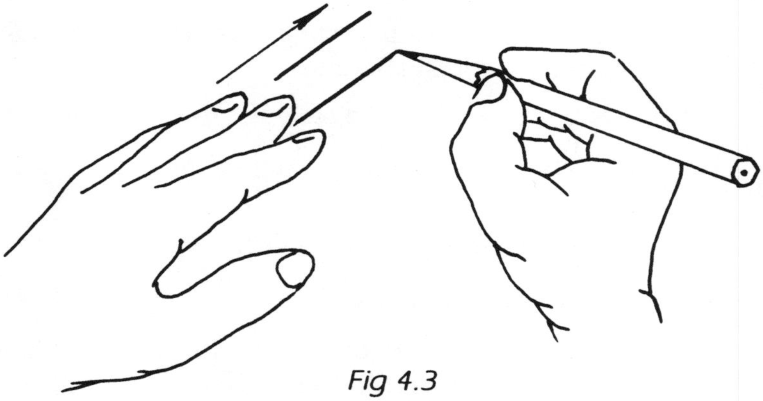

Fig 4.3

Activity 1 **Straight-line exercises**

Practise these straight-line exercises to improve your sketching motions. Certain movements with the pencil are easier to control in certain directions than others. Practise fig 4.4a–g several times on rough-work paper. By practising these straight lines you will learn to handle your pencil more smoothly. Remember to turn the paper around to the most comfortable drawing position.

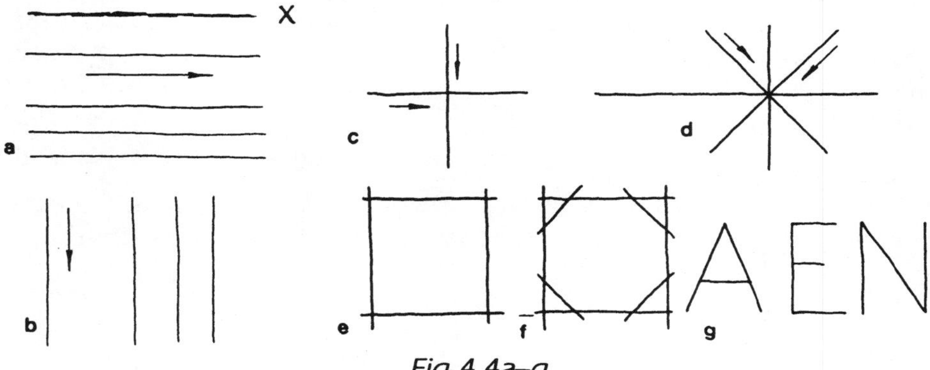

Fig 4.4a–g

3.4 Sketching curved figures and geometric shapes

Objects which look difficult to sketch can be visualised as being composed of a few simple geometric forms. These forms are usually squares, rectangles, right-angled triangles, circles, ellipses, hexagons (six-sided figures) and octagons (eight-sided figures).

Often, when sketching a part, you need to draw both straight and curved lines.

Construction of a freehand arc (also called a fillet)

There are five basic steps (see fig 4.5) when drawing an arc which joins two straight lines.

Step 1: Extend the straight lines so that they intersect.

Step 2: Mark off the same distance from the intersection on both lines.

Step 3: Draw a diagonal line through these points.

Step 4: Place a dot approximately in the **centre of the triangle**.

Step 5: Start at one of the lines and sketch an arc which runs through the dot and ends on the other line. Darken the arc and erase all **unnecessary** lines.

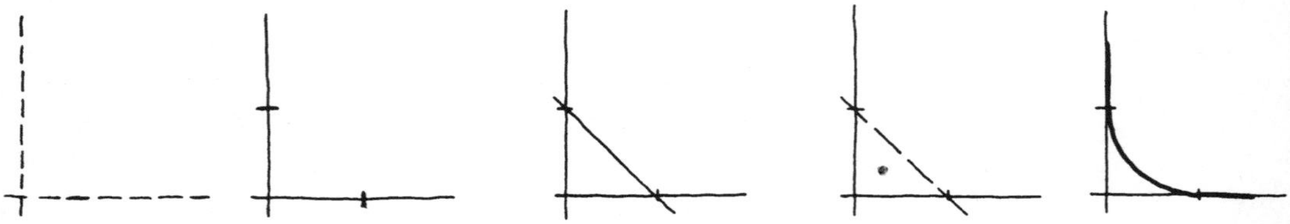

Step 1: Two lines intersect | *Step 2: Mark off equal distances* | *Step 3: Draw a diagonal* | *Step 4: Locate centre dot of triangle* | *Step 5: Sketch arc*

Fig 4.5

Sketching arcs connecting straight lines and curved lines

To draw an arc between a straight line and a curved line, follow the same construction as in the example above.

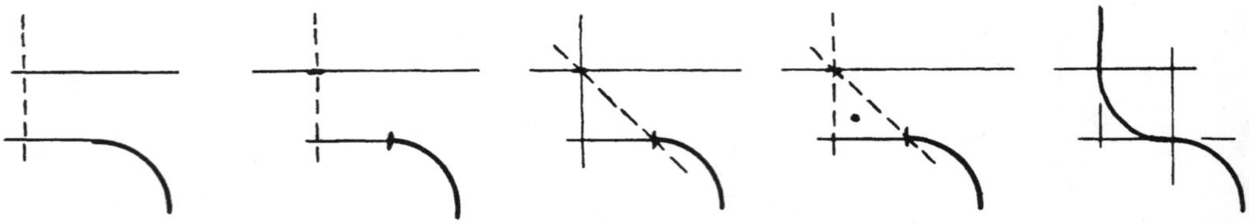

Step 1: Intersect lines | *Step 2: Mark off points* | *Step 3: Form triangles* | *Step 4: Locate triangle centre* | *Step 5: Sketch arc*

Fig 4.6

To draw an arc between two curved lines, use the same construction as for the two previous examples.

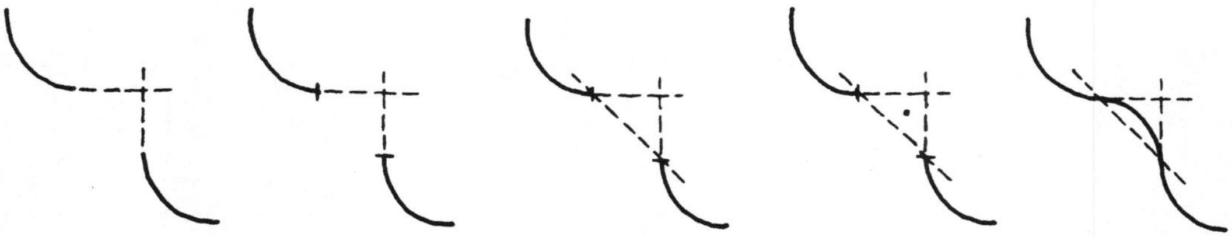

Step 1: Intersect lines | *Step 2: Mark off points* | *Step 3: Form triangles* | *Step 4: Locate triangle centre* | *Step 5: Sketch arcs*

Fig 4.7

Steps in sketching a square

Step 1: Sketch the centre of the square first.

Step 2: Mark off the approximate distances on the two centre lines.

Step 3: Draw the square with very feint lines to see if your proportions are correct.

Step 4: Draw the outlines with heavy lines if you are satisfied with the proportions. Erase all the unnecessary lines.

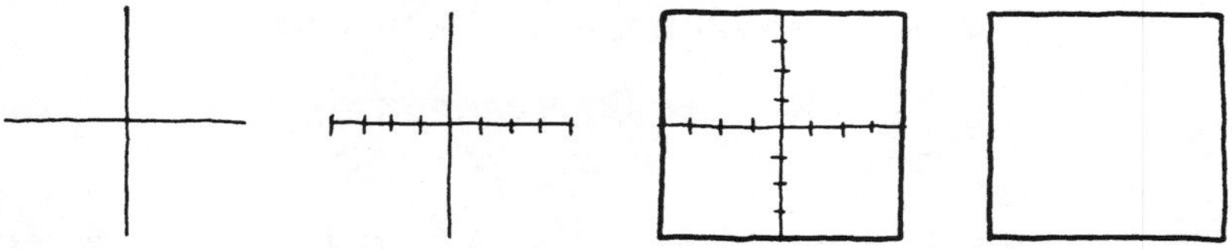

Fig 4.8

Steps in sketching a rectangle

Follow the same steps as in the above example of how to draw a square.

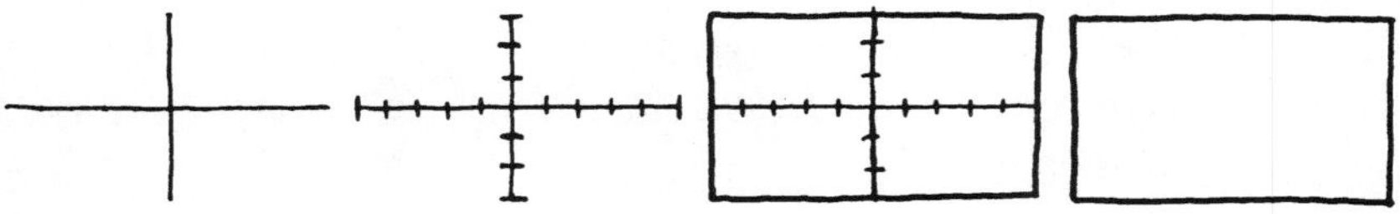

Fig 4.9

Steps in sketching a right angled triangle

Follow the same steps as in the two previous examples, but use only half the centre lines.

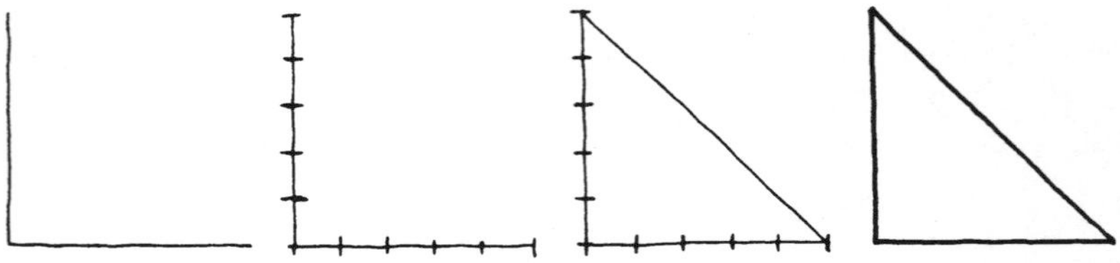

Fig 4.10

Sketching circles

There is more than one method for drawing a circle. In this unit, two methods are shown. Make sure that you understand the first one because it is based on the method of how to sketch **arcs** or **fillets**. Remember, a circle has a diameter and a radius (the radius is half the size of the diameter).

Method 1

This method is based on the method used for arcs.

Step 1: Draw the horizontal and vertical centre lines.

Step 2: Mark off the approximate radius of the circle from the intersection point of the two centre lines.

Step 3: Draw lines connecting at the ends of the centre lines and estimate the centres of the triangles.

Step 4: Sketch arcs to form the complete circle.

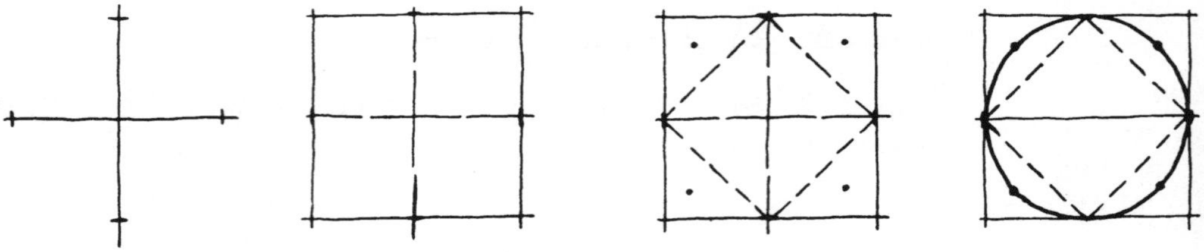

Step 1: Laying out diameter | *Step 2: Drawing square* | *Step 3: Drawing diagonals and triangle centres* | *Step 4: Sketch arcs*

Fig 4.11

Method 2

Step 1: Draw the centre lines (horizontal and vertical).

Step 2: Draw two extra lines making an angle of $\pm 45°$ with the other two centre lines.

Step 3: Mark off the radii on all the lines.

Step 4: Draw arcs at the end of these lines and join them to form the outline of the circle.

Remember always to darken the outline of the circle and erase all the guide lines to simplify the reading of the sketch.

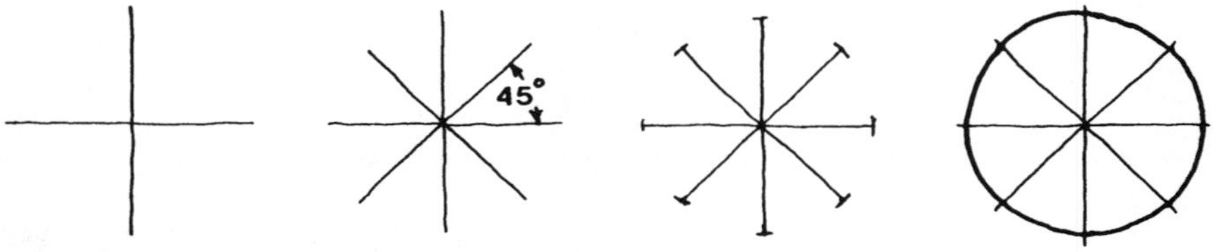

Step 1 *Step 2* *Step 3* *Step 4*

Fig 4.12

Sketching a circumscribed hexagon

Step 1: Follow method 1 (see page 64) to sketch a circle.

Step 2: Draw lines tangent to the circle at 60° as shown in fig 4.13.

Step 3: Darken the hexagon formed by the tangent lines and erase all unnecessary lines.

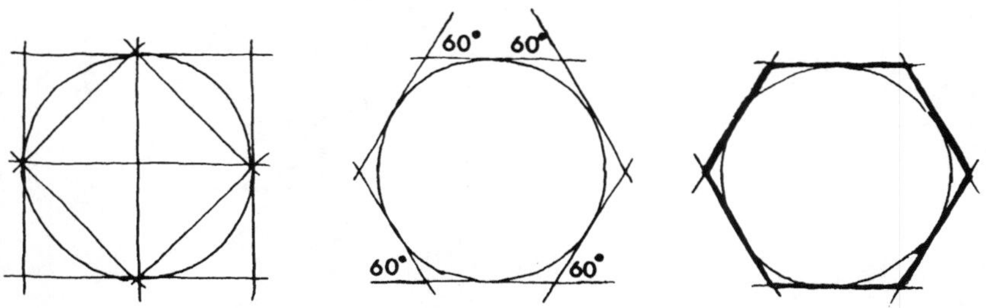

Fig 4.13

Sketching an inscribed hexagon

Draw the centre lines and two extra lines at ±30° to the horizontal, and mark off the radii. Draw arcs at the ends of these lines and join them to form a circle. Draw the hexagon to connect the points of intersection.

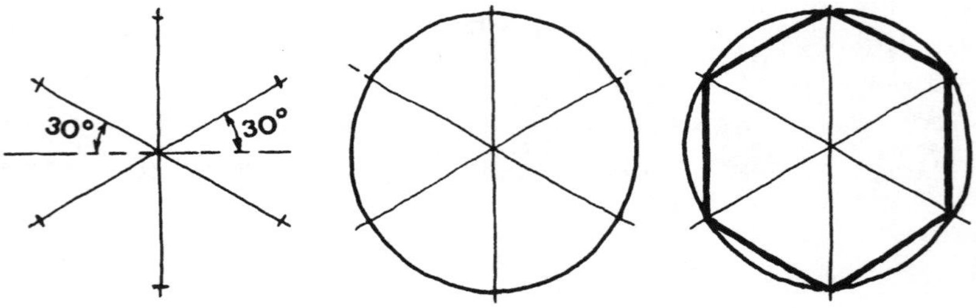

Fig 4.14

Sketching a circumscribed octagon

Follow method 2 (see page 65) for sketching a circle and complete the circumscribed octagon as shown.

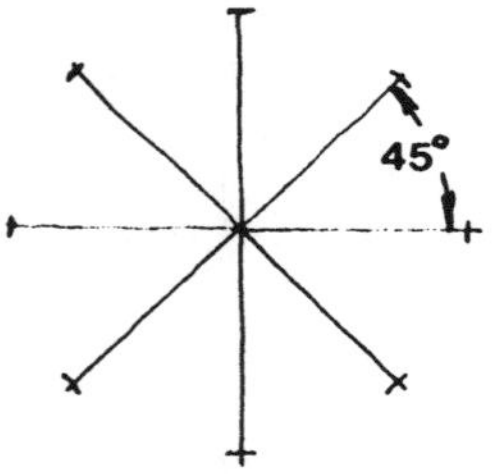

Step 1

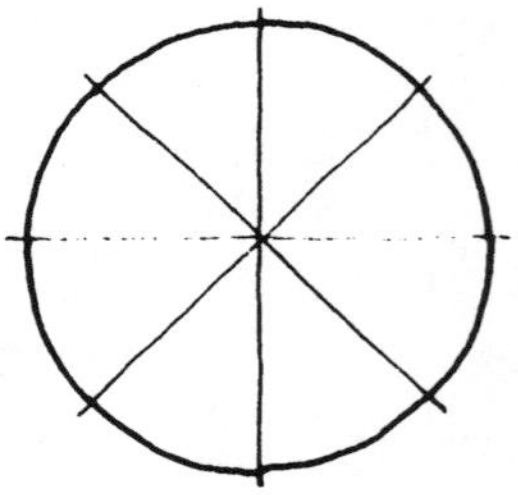

Step 2

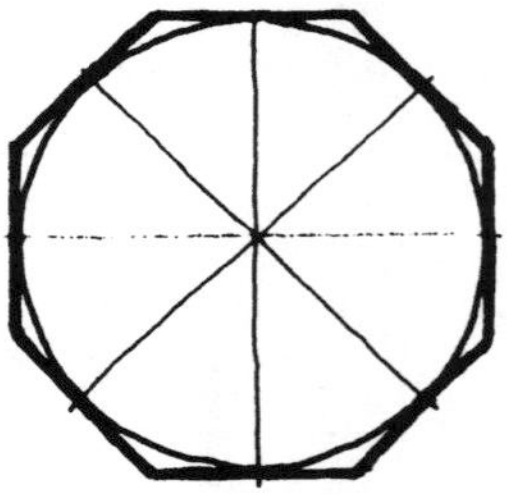

Step 3

Fig 4.15

Sketching an ellipse

An ellipse is a circle that you see at an angle. In other words, the one centre line is longer than the other one. The longer centre line is called the **major axis** and the shorter one, the **minor axis**. This will be discussed in more detail in the next unit.

To sketch an ellipse, draw a rectangle first by following the steps described on page 63. Then fill in the arcs as shown.

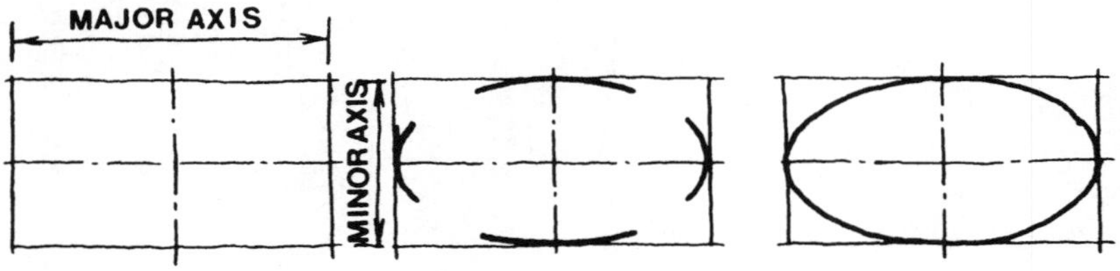

Step 1 *Step 2* *Step 3*

Fig 4.16

Activity 2

Curved-line exercise

Practise this exercise several times to improve your sketching motions. Certain movements are easier to control in certain directions than in others; you should therefore **rotate the sketching paper around when sketching**.

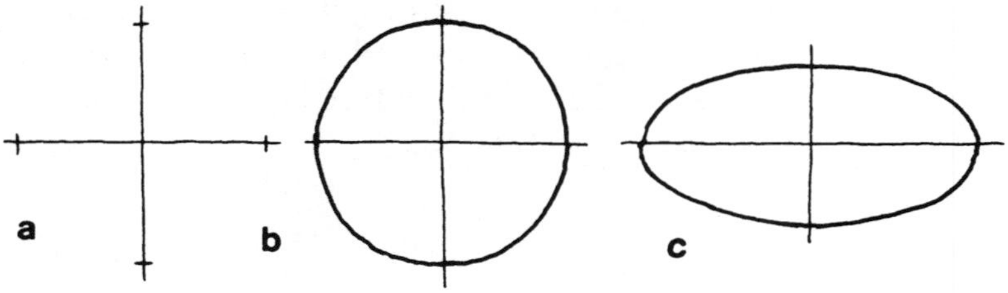

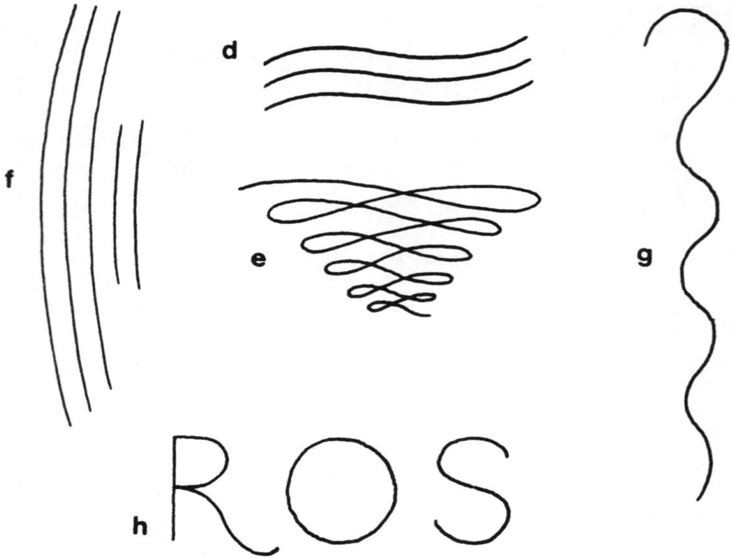

Fig 4.17a–h

3.5 Sketching irregular shapes

At this point it is important to remind you that all the techniques described in this unit have been thoroughly **tested** and have been found to be essential in training the beginner to sketch accurately.

All the steps for drawing irregular shapes are based on the methods explained above and you must therefore be able to draw all the above lines and figures before you attempt to sketch irregular shapes.

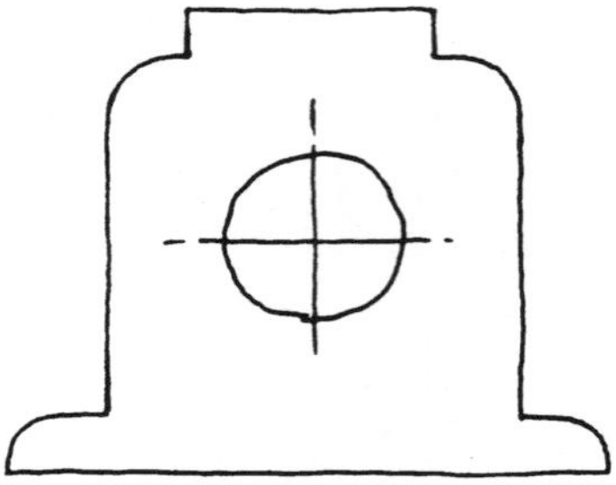

Fig 4.18 Shaft support

Step 1: Lay out the two basic rectangles required for the shaft support. Use light lines and draw the outline of the desired overall sketch.

Step 2: Draw the centre lines (horizontal and vertical).

Step 3: From the horizontal centre line, mark off the square for the arc at the top (both sides). Remove unnecessary lines to simplify the reading of the sketch.

Step 4: Draw the squares for the round corners at the top and the base, and also for the centre hole.

- **a** Draw the diagonals and form the triangles.
- **b** Locate dots in the centres of the triangles through which each arc will pass.

Step 5: Draw the arcs and circle. Start at one diagonal and draw the arc through the dot to the next diagonal. Continue until the circle is completed.

Reproduced by permission Basic Blueprint Reading and Sketching. By Olivio. Delmar Publishers, Albany, New York Copyright 1999

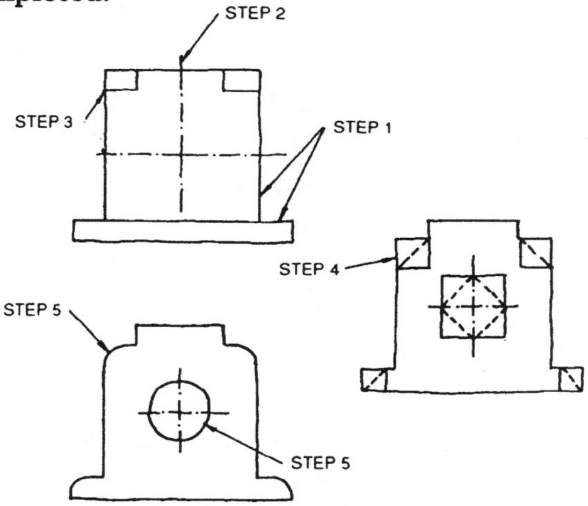

Fig 4.19 Sketches of the steps

To complete the sketch, darken all outlines. Erase those lines used in the construction that either do not simplify the sketch or are not required to interpret the sketch quickly and accurately. Insert the dimensions that you measured.

Fig 4.20b shows the three steps for making a freehand sketch of the hexagonal nut shown in fig 4.20a.

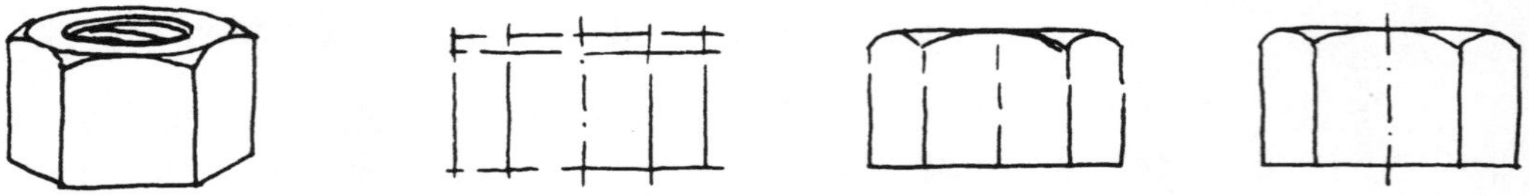

Step 1 *Step 2* *Step 3*

a *b*

Fig 4.20a–b

Fig 4.21b shows the three steps for making a freehand sketch of the cover shown in fig 4.21a.

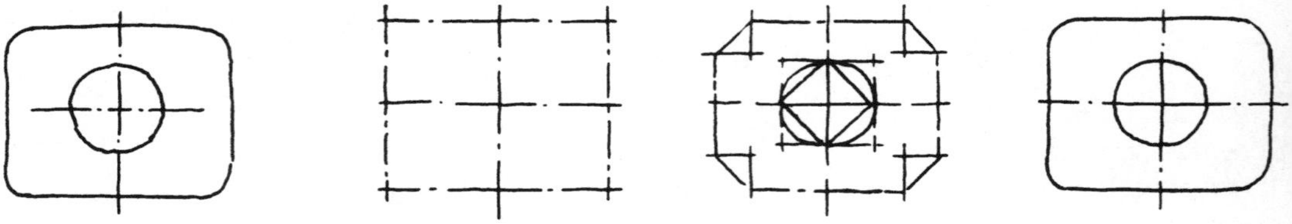

Step 1 *Step 2* *Step 3*

a *b*

Fig 4.21a–b

3.6 Isometric sketching

Isometric and oblique sketches are the two most common forms of pictorial drawing in which two or more surfaces are shown in one view.

The isometric sketch revolves around three major lines called isometric axes or baselines (see fig 4.22).

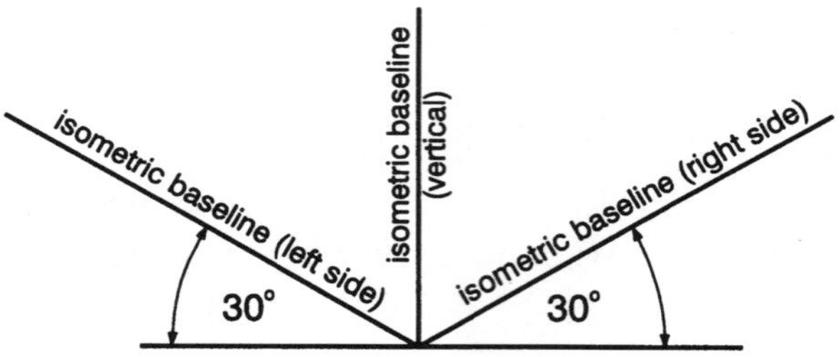

Fig 4.22

Making a simple isometric sketch (horizontal and vertical lines)

An isometric object is sketched by positioning it so that the object seems to rest on one corner.

The basic four steps for drawing a rectangular steel block with the following dimensions are set out below:

Width = 20 mm Length = 30 mm Thickness = 10 mm

Step 1: Sketch the three isometric axes as shown. Note the angles!

Step 2: Lay out the width: 20 mm along the left axis line; the length: 30 mm on the right axis line; and the thickness: 10 mm on the vertical axis line.

Step 3: Draw lines from these layout points parallel to the three axes.

Step 4: Darken the outline, erase unnecessary lines and dimension the sketch.

In isometric sketches, the dimensions are placed parallel to the edges.

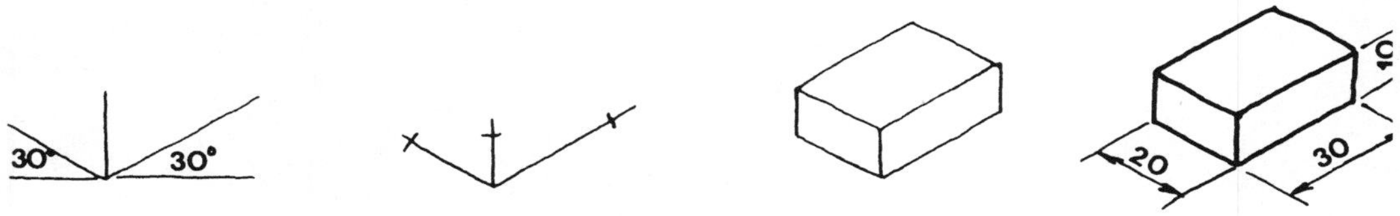

Step 1 *Step 2* *Step 3* *Step 4*

Fig 4.23

Sketching slant lines isometrically

Those lines that are parallel to the axis may be measured in their true lengths. Slant lines representing inclined surfaces are not shown in their true lengths in isometric sketches. In most cases, the slant lines for an object may be drawn as follows:

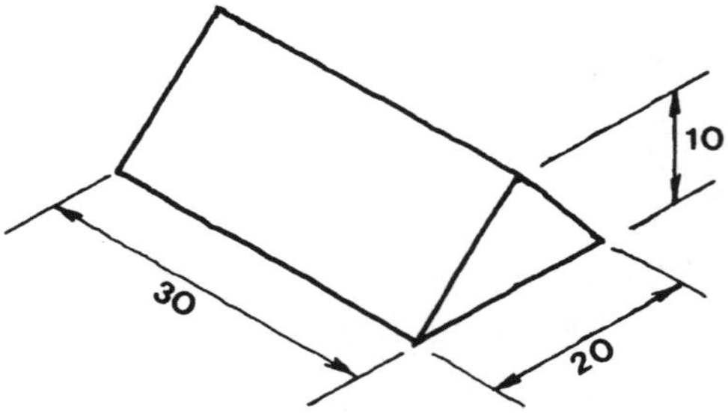

Fig 4.24 Part to be drawn isometrically

Step 1: Draw the three major isometric axes.

Step 2: Measure distance AB on the left axis and distance AC on the right axis.

Step 3: Measure distance AD on the right axis and draw a vertical line from this point. Lay out distance DE on this line.

Step 4: Draw triangle ACE.

Step 5: Draw line EF parallel to AB and BF parallel to AE.

Step 6: Darken the outline and erase unnecessary lines.

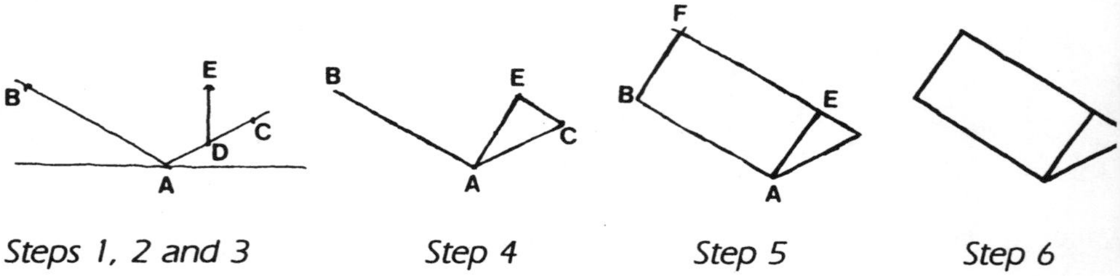

Fig 4.25

Sketching isometric circles and arcs

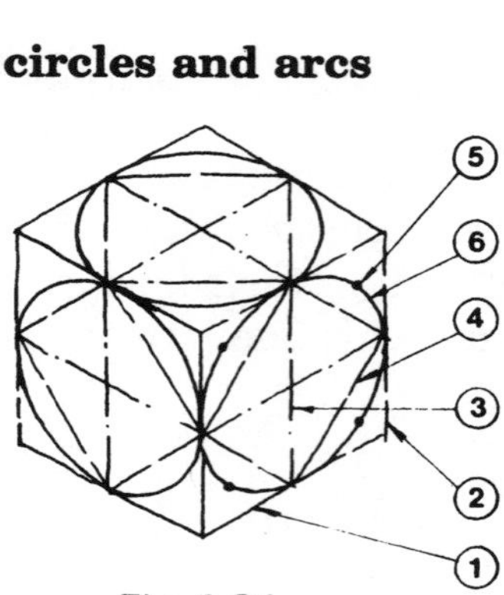

Fig 4.26

Application of circles and arcs isometrically

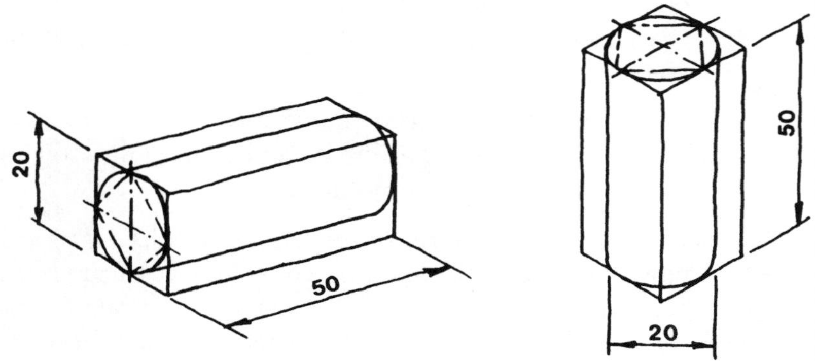

a *b*

Fig 4.27a–b Isometric sketch of a round bar

Since all the circles are not the true size isometrically, they will be elliptical in shape on all the planes.

3.7 Oblique sketching

An oblique sketch is another form of pictorial drawing where two or more surfaces are shown at one time on one drawing.

The difference between **isometric sketches** and **oblique sketches** is the axes, as shown in fig 4.28a–b.

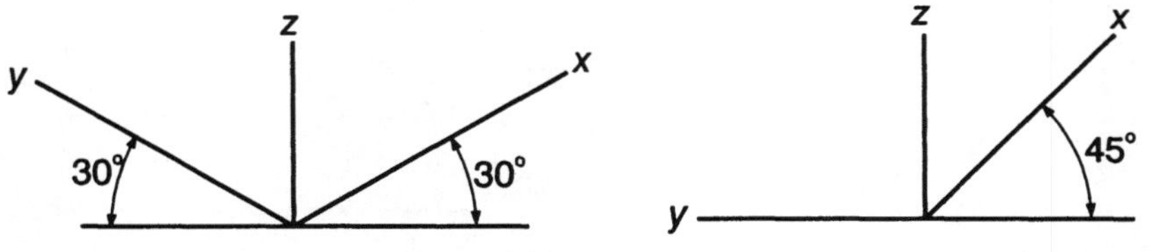

a Isometric *b Oblique*

Fig 4.28a–b

Making oblique sketches with straight lines

The steps in making oblique sketches are very simple. For example, to draw an oblique sketch of a rectangular block, follow the seven basic steps below:

Step 1: Select the view of the object which gives most of the desired information.

Step 2: Draw light horizontal and vertical baselines.

Step 3: Lay out the edges of the block in the front view from the baselines. All the vertical and horizontal lines must be parallel to the baselines. (All lines will be in their true size and shape in this view.)

Step 4: Start at the intersection of the vertical and horizontal baselines, and draw a line at an angle of 45° to the baseline. This line is called the oblique baseline.

Step 5: Draw the remaining lines for the right side and top view parallel to either the oblique baseline or to the horizontal or vertical baselines. Since these lines are not in their true size or shape, they should be drawn in the correct proportion to the front view.

Step 6: Place the dimensions so that they are parallel to the baselines.

Step 7: Erase any unnecessary lines. Darken object lines to make the sketch easier to interpret.

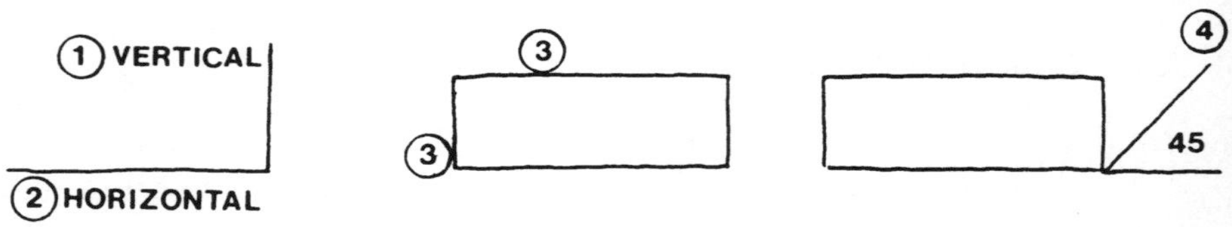

Steps 1 and 2 *Step 3* *Step 4*

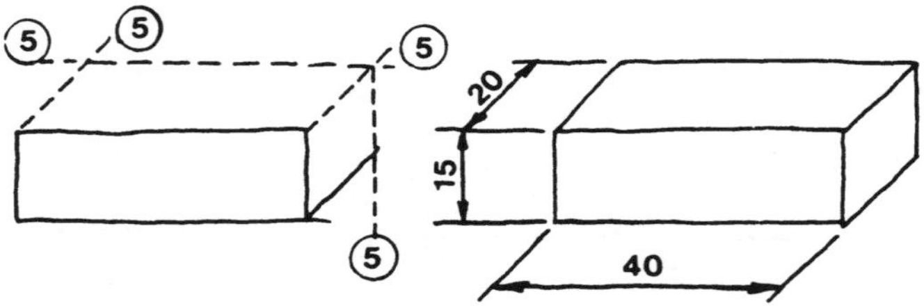

Step 5 *Step 6*

Fig 4.29

Sketching circles in oblique

A circle or arc located on the front view of an oblique sketch is drawn to true size and shape. However, since the top and side views are distorted, a circle or arc will be elliptical in these views.

Step 1: Draw horizontal, vertical and oblique baselines.

Step 2: Sketch lines parallel to these baselines or axes to form a cube.

Step 3: Draw centre lines and diagonals in the front, right side and top faces.

Step 4: Locate the midpoints of triangles formed in the three faces.

Refer back to the section 3.4 on sketching arcs and circles. We cannot stress enough the importance of practising these different sketches until you can draw them neatly and accurately.

Step 5: Draw curved lines through these points.

The circle appears in its true size and shape in the front plane, and as an ellipse in the right and top planes.

Step 6: Touch up and darken the curved lines. Erase unnecessary guide lines if they are of no value in reading the sketch.

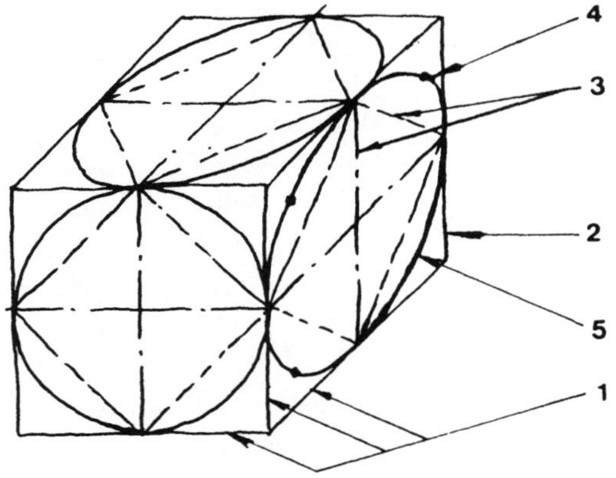

Fig 4.30

Right and left oblique sketches

Up to now we have viewed objects from the right side. In many cases, the left side of an object must be sketched because it may contain better details.

The same principles and techniques for making oblique sketches apply, regardless of whether the object is drawn in the right or left position.

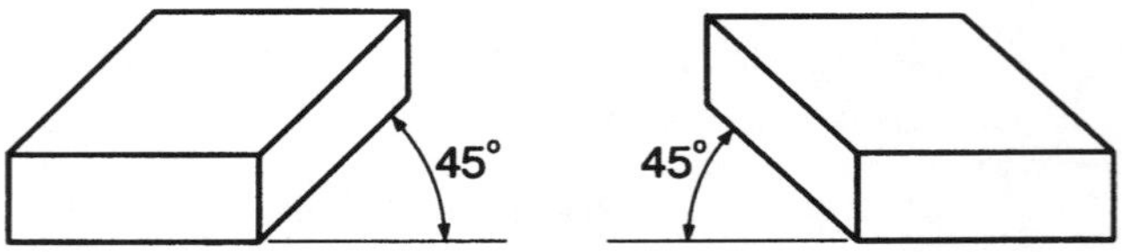

a Right oblique *b Left oblique*

Fig 4.31a–b

Foreshortening

When a line in the side and top views is drawn in the same proportion as lines in the front view, the object may appear to be distorted and longer than it actually is. To correct the distortion, the lines are drawn shorter than the actual size so that the sketch looks balanced. This draughting technique is called **foreshortening** (see fig 4.32).

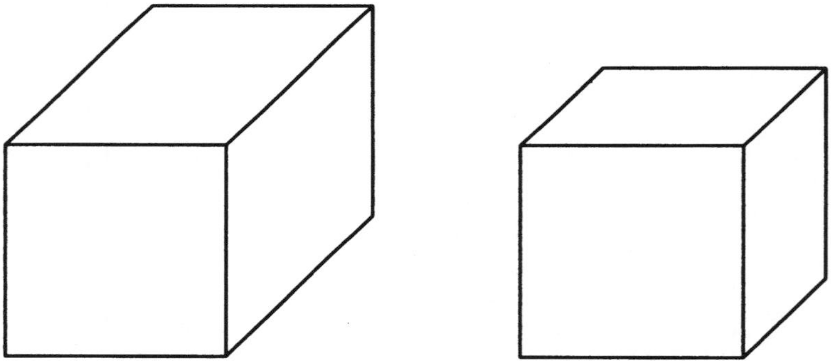

TRUE CUBE FORESHORTENED CUBE

Fig 4.32

Activity 3

Procedure

1 Sketch the chisel shown in fig 4.33a using the following steps, as illustrated in fig 4.33b.

- Sketch the centre lines in position and mark out the proportions by means of light enclosing boxes.
- Using light lines, build up the outline and inside detail.

- Use an eraser to clean the sketch and then make the outline and detail clear.

2 Sketch the hexagonal nut shown in fig 4.33c using the steps illustrated in fig 4.33d.

- H shows three steps in making a freehand sketch of a cover, shown in 'G'.

3 Sketch the cover shown in fig 4.33e using the steps illustrated in fig 4.33f.

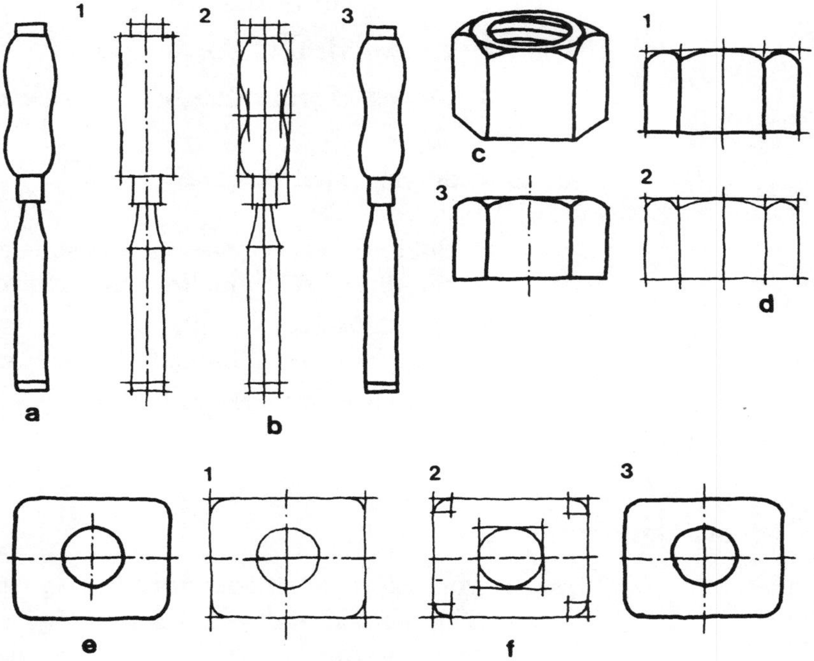

Fig 4.33a–f

4 Summary

To be able to make neat, well-proportioned freehand sketches is as important as the ability to make drawings at a drawing-board using a T-square and instruments. Many people will say that freehand sketching is, in fact, more important because workers and engineers, not otherwise employed in the drawing office, frequently find it necessary to express their ideas by means of sketches.

A word of caution – do not expect perfection in your first efforts. The desired skill will be more readily acquired by having a definite method to follow. The most important thing to aim for is good proportion. No matter how good the linework, if the proportions are poor the sketch will be poor as well.

Remember to use a B or HB pencil, and practise regularly on a rough-work pad.

Pay particular attention to the following points:

1 The layout of the **views**
2 **Proportion** in the sketches
3 **Types of pencils** used for freehand sketching

In addition you must be able to:

1 Sketch all the **types of lines**, such as horizontal, vertical, slant, straight lines and curved figures
2 Construct arcs, circles, rectangles, right angles, hexagons, octagonal ellipses and irregular figures
3 Sketch **isometric** and **oblique** objects

Self-evaluation

Before you attempt the questions below, you must make sure that you can apply the skills and knowledge contained in this unit. These exercises for self-evaluation are typical of the assignment questions a lecturer would set for you.

Draw the following objects **freehand**.

One dimension is given on each object. Draw the object in good proportion to dimension (scale: full size).

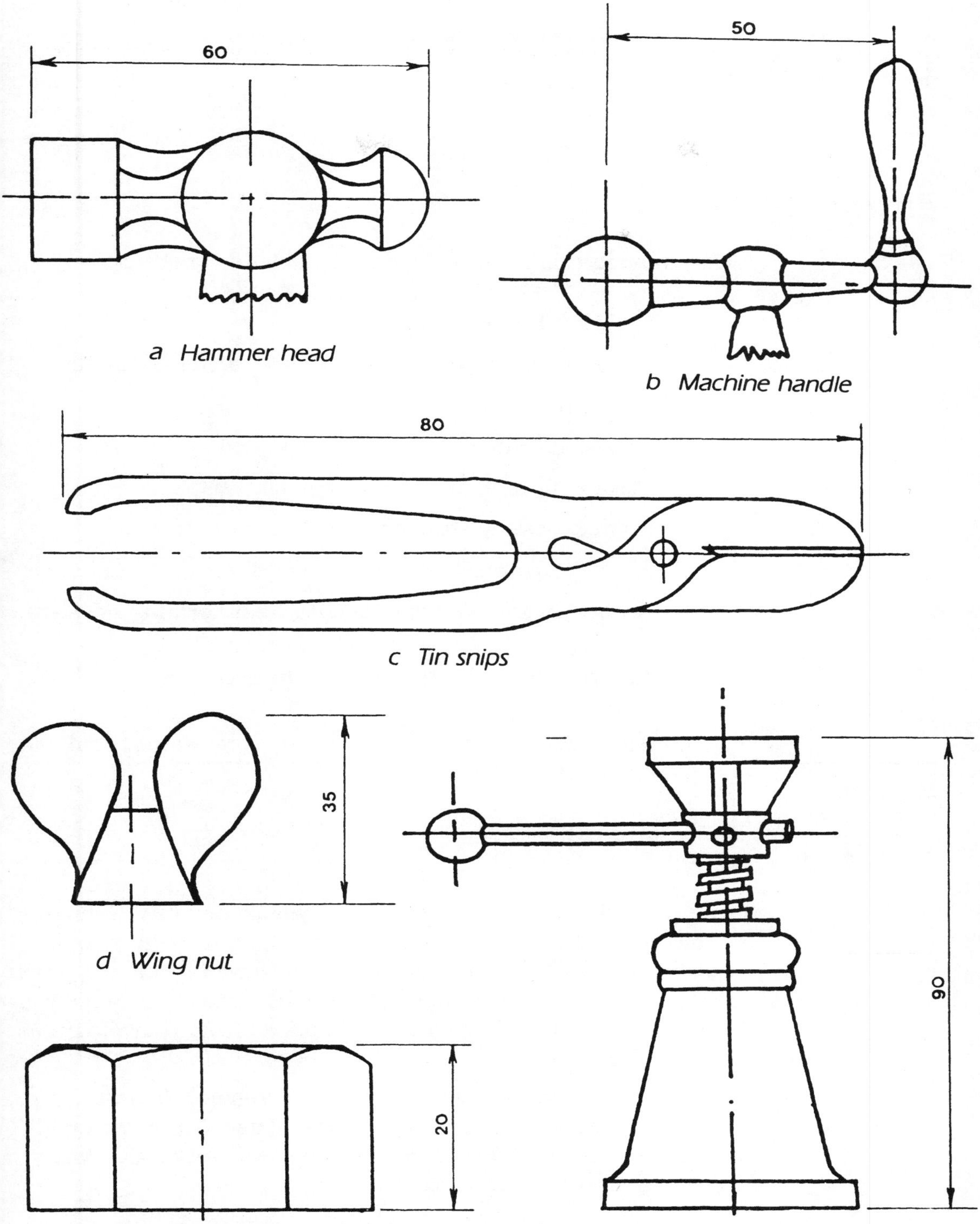

a Hammer head

b Machine handle

c Tin snips

d Wing nut

e Hexagon nut

f Lifting jack

Fig 4.34a–f

Unit 5

Constructions

Study objectives

After studying this unit, you should be able to construct:

- ◇ A bisected line
- ◇ A perpendicular to a line at a given point on the line
- ◇ A bisected angle
- ◇ A perpendicular at the end of a line
- ◇ Parallel lines
- ◇ A circle passing through any three given points
- ◇ An arc of a circle that touches lines forming a right angle
- ◇ An arc of a circle that touches lines forming an acute angle
- ◇ An arc of a circle that touches lines forming an obtuse angle
- ◇ An ellipse, using both the four-centre method and the trammel method.

1 Introduction

This unit introduces you to a special type of drawing technique in which the solution to a problem is obtained through graphical methods. This is different from what you have been doing until now. It is important that your drawings be accurate because you will have to measure the lengths of the lines from your diagrams to obtain the answer.

Remember that you **never** scale measurement lengths from **working drawings**; you should **always use the printed dimensions**. The diagrams in this unit are **not** working drawings, they are geometric constructions.

In this unit you will be introduced to all the different constructions. In each case the constructions must be done carefully in order for the results to be accurate.

2 Geometrical constructions

2.1 *Bisecting lines and angles*

What do we mean when we speak of bisecting a line or an angle?

When you bisect a line or an angle, you are dividing the line or angle into two equal parts. This is usually done by means of a compass.

How to bisect a line with the aid of a compass

Step 1: Draw a line AB of any given length.

Step 2: Set your compass to a radius of more than half of the length of line AB. Draw two arcs with the point of the compass positioned at A, one above and one below line AB. With the same radius on the compass, draw another pair of arcs from point B.

Step 3: The point where the two arcs meet at the top is point C and the point where the two arcs meet below the line is point D. Join C and D. CD will bisect AB at right angles at O. AO is therefore equal to OB.

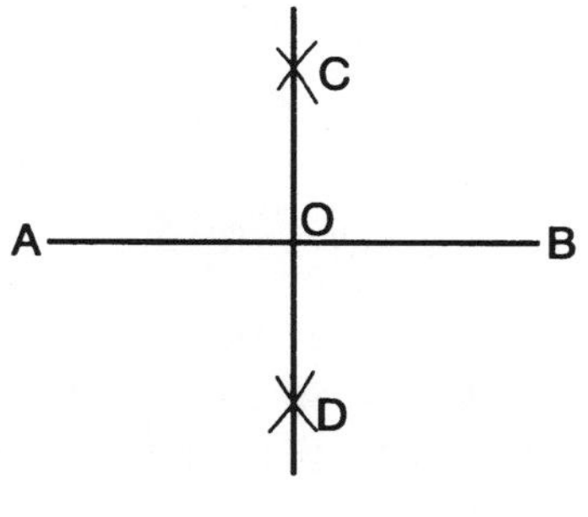

Fig 5.1

Activity 1 Draw various slanted lines of different lengths and divide them into two equal parts by bisecting them.

How to bisect any given angle with the aid of a compass

Step 1: Adjust your compass to any radius. With the point of your compass positioned at B, draw an arc cutting AB and CB at D and E respectively.

Step 2: With the point of your compass positioned at D, draw an arc again. Then using the same radius, draw another arc from point E. The point where the two arcs cut or intersect is called F.

Step 3: Join BF which bisects angle ABC. Line BF is the **bisector** of angle ABC.

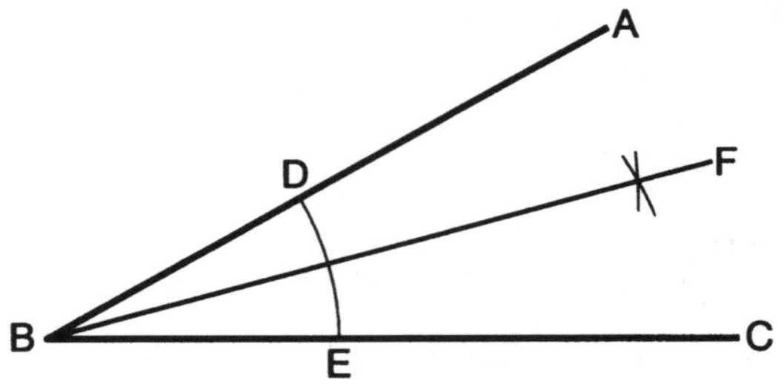

Fig 5.2

Activity 2

Draw different angles and, by following the steps outlined above, find the bisector of each of the angles.

2.2 Perpendiculars

What is a perpendicular?

A perpendicular occurs when one line intersects another line at 90°. The two lines will be perpendicular to each other and are referred to as perpendiculars.

How to construct a line perpendicular to another line at a given point, using your compass (in this case C is the given point)

Step 1: From point C on the line AB, and with a suitable radius, draw arcs intersecting AB at D and E.

Step 2: Position the point of your compass at D and draw an arc above C and, using the same radius, draw another arc from point E. The two arcs will intersect at F.

Step 3: Join FC and this line will be perpendicular to line AB at point C.

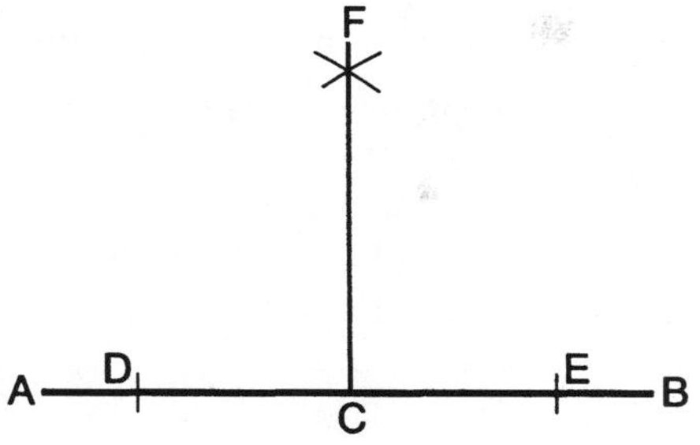

Fig 5.3

Activity 3 Draw several lines and construct a perpendicular at any point on those lines.

How to construct a perpendicular at the end of a line

Step 1: Draw a line AB.

Step 2: Choose any point outside line AB, which you will call C.

Step 3: With the compass point on point C and using the radius CB, draw an arc to intersect AB at D.

Step 4: Join DC and extend to intersect the arc at E.

Step 5: Join EB to form a perpendicular.

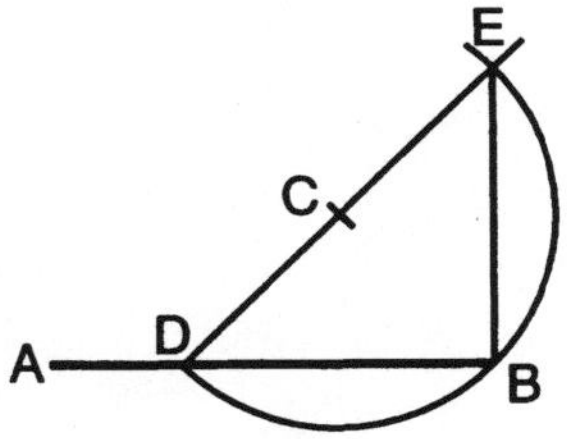

Fig 5.4

Activity 4 Practise constructing perpendiculars at the end of a few lines slanted at different angles.

2.3 Parallel lines

What are parallel lines?

We can compare parallel lines to railway lines or telephone wires. Parallel lines lie next to each other, the same distance apart, and never meet.

How to construct a line parallel to line AB at a given distance, using a ruler and compass

Step 1: Draw a straight line AB. Mark any two points on AB as point C and point D. Set the required distance on the compass and, using the same radius, strike arcs E and F from points C and D respectively.

Step 2: Draw a **tangent** line to touch arcs E and F. This new line will be parallel to AB.

The term 'tangent' refers to a straight line that touches an arc of a circle only once.

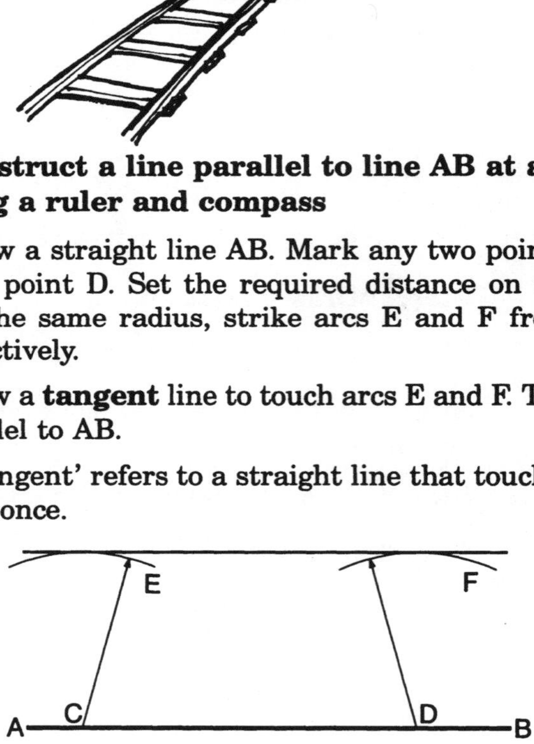

Fig 5.5

Activity 5

Practise constructing parallel lines, as shown above. It is important that you know how to do this, because in later exercises this type of construction will be used frequently.

How to construct a circle so that it passes through any three points A, B and C, which are not on a straight line

Step 1: Given points A, B and C, join the three points to form two lines AB and BC.

Step 2: Bisect AB and BC, i.e. draw perpendicular bisectors, and extend them until they intersect. The point where they intersect will be the centre of the circle, point O.

Step 3: With radius OA, draw the circle which will also pass through points B and C.

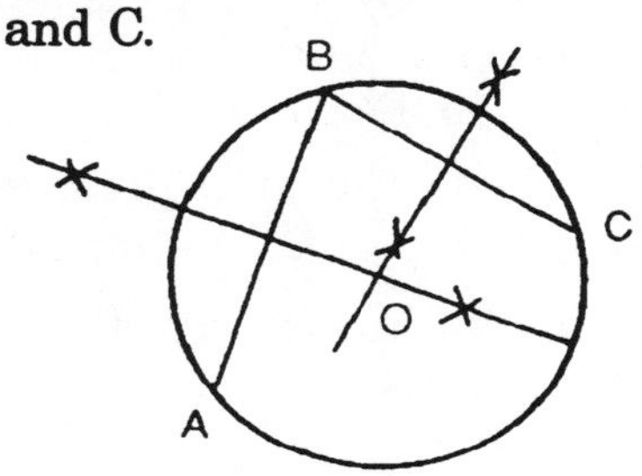

Fig 5.6

How to draw parallel lines with the use of set squares

Step 1: Draw line AB, as shown.

Step 2: Place one set square on line AB.

Step 3: Place the other set square against the set square on line AB, as shown, and hold it firmly against the paper to prevent it from moving.

Step 4: Slide the set square on line AB along the other set square, as shown; in this way you can construct several parallel lines to AB.

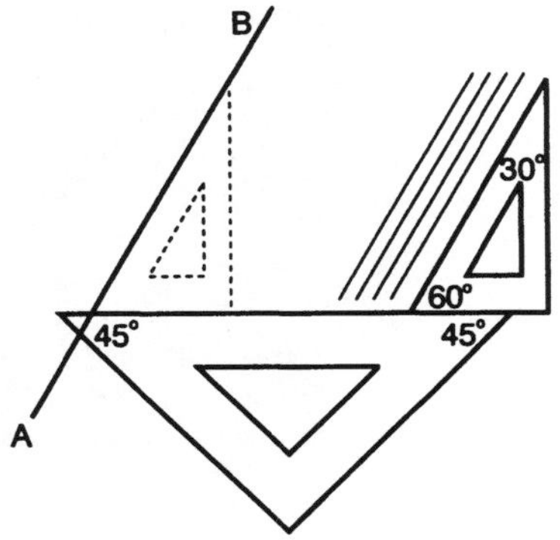

Fig 5.7

How to divide a line into equal parts

Step 1: Draw a line of any length and mark off a given length AB.

Step 2: From point A, draw an inclined line using your set square. Use any angle (30°, 45° or 60°).

Step 3: Mark off seven consecutive segments on the inclined line with your compass. You can use any radius, as long as you use the same radius for each segment that you mark off. The seventh segment is marked point F.

Step 4: Now join point B to point F, the last segment on the inclined line.

Step 5: Use the set square method to draw lines from all the other points parallel to line BF.

Step 6: The lines parallel to BF divide AB into seven equal segments or parts.

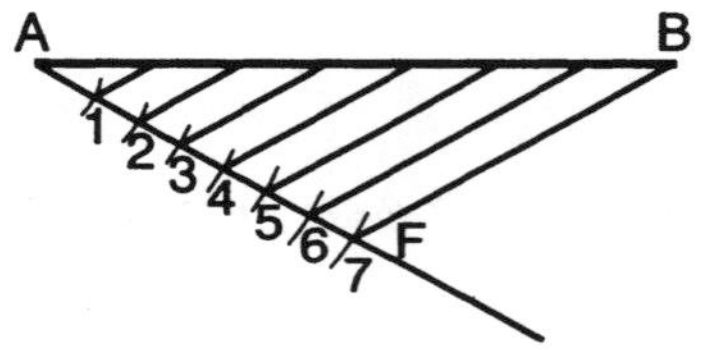

Fig 5.8

2.4 Setting out angles with the aid of set squares

Although the set squares are normally used to set out angles of 90°, 60°, 45° and 30°, they can also be used to set out a number of other angles. See the following examples:

The 45° set square is used to set out an angle of 135°.

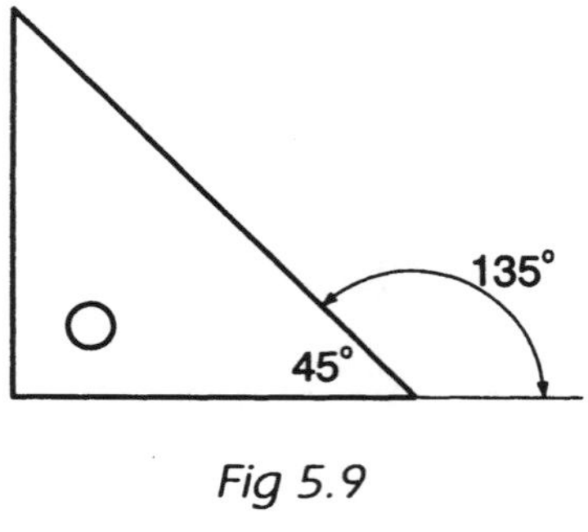

Fig 5.9

The 60°–30° set square is used to set out an angle of 150°.

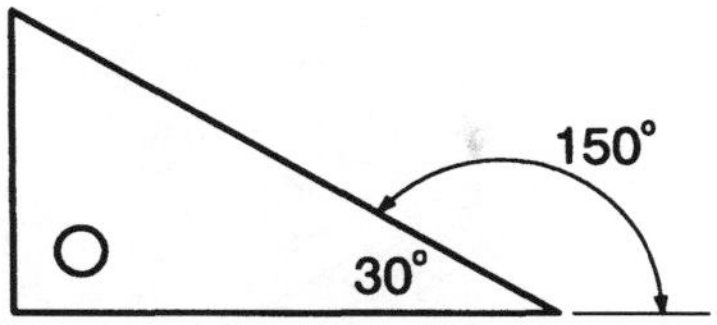

Fig 5.10

The 60°–30° set square is used to set out an angle of 120°.

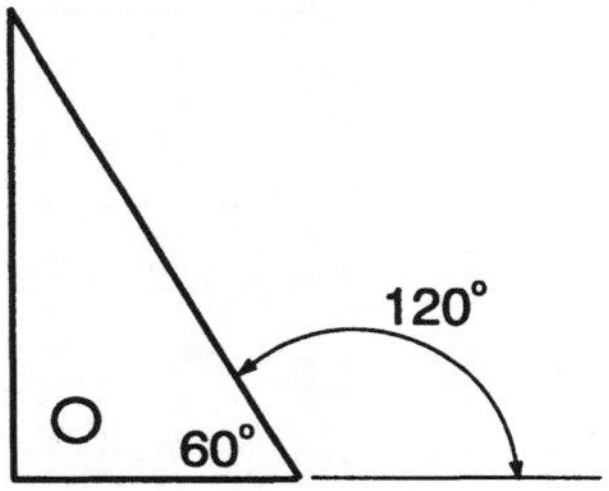

Fig 5.11

Two set squares are used in conjunction to set out an angle of 15°.

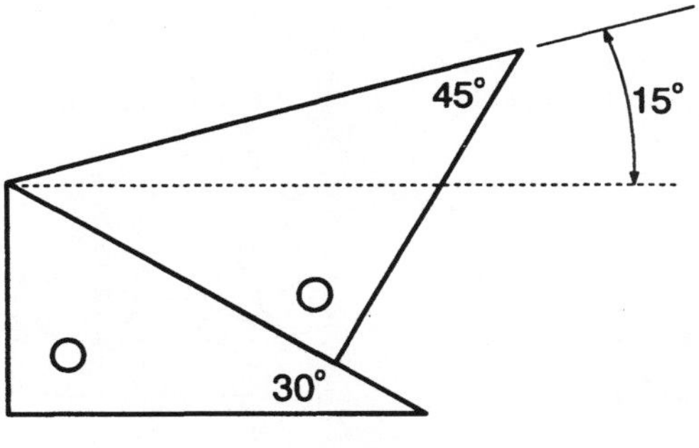

Fig 5.12

Two set squares are used in conjunction to set out an angle of 75°.

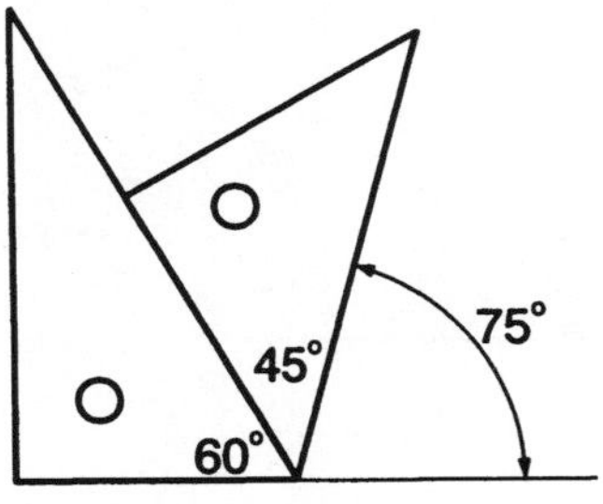

Fig 5.13

Two set squares are used in conjunction to set out an angle of 105°.

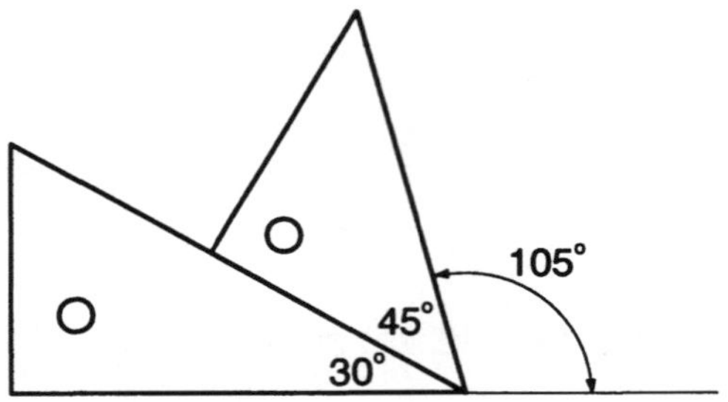

Fig 5.14

Activity 6

1 Draw a horizontal line 95 mm long and bisect it.

2 Draw a 30° line 100 mm long and bisect it.

3 Draw a 75° angle and bisect it.

4 Draw a 105° angle and bisect it.

5 With the aid of set squares, accurately draw fig 5.15a–e full scale. Insert the dimensions, as shown in unit 3.

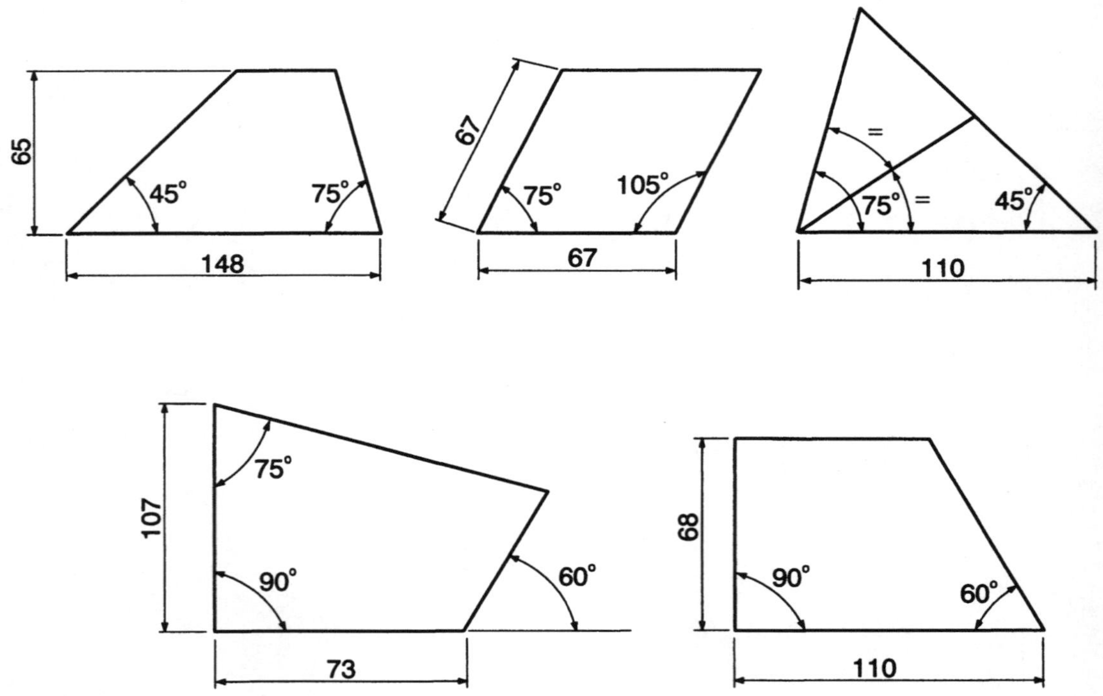

Fig 5.15a–e

2.5 Hexagons (Six-sided figures)

Although we will be concentrating on hexagons and octagons in this course, there are several different terms and shapes you need to be familiar with to do geometrical constructions. A table containing these different geometric terms has been included at the end of this unit for easy reference.

How to construct a regular hexagon with a given side

Step 1: Draw line AB to the given length.

Step 2: Using a 60° set square, project lines upwards from point A and B respectively, as shown. Mark off AF and BC to the same length as AB.

Step 3: From points F and C, project with the set square to points E and D respectively, but first mark off the side measurements to the given size.

Step 4: Complete the hexagon by drawing line ED.

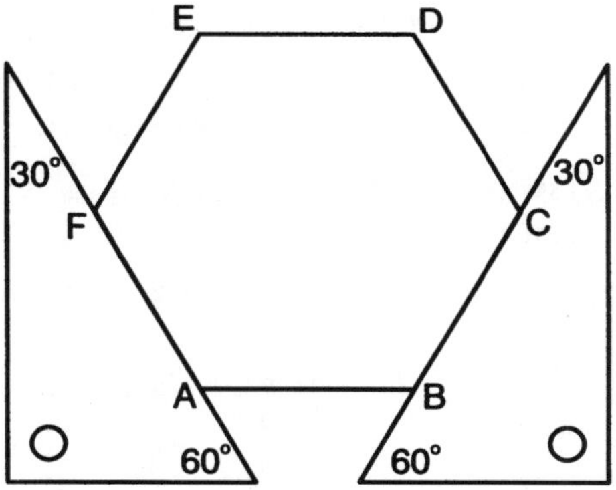

Fig 5.16

How to construct a regular hexagon with a given size across the flats

Step 1: Mark off the given size BF.

Step 2: By using the set square turned around to the 30° side, project lines downward from points B and F to locate point A.

Step 3: From points B and F, project vertical lines to points C and E. (FE and BC are perpendicular to the edge of the T-square.)

Step 4: Complete the hexagon by projecting at 30° from points C and E to the centre D.

All the sides of a hexagon are equal. BF is the distance across the flats and is indicated by the symbol A/F.

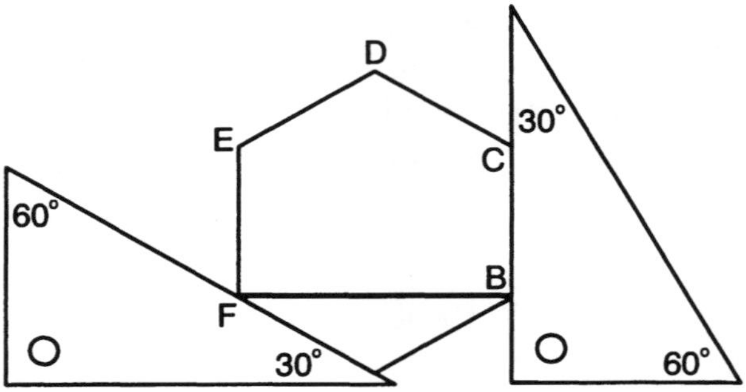

Fig 5.17

How to construct a regular hexagon by means of the circumscribed circle method

Step 1: Draw line MN and mark off AB to the given size.

Step 2: Draw an arc from points A and B, with a radius equal to AB. Where the two arcs intersect, you obtain point O.

Step 3: With centre O and radius OA, draw the circumscribed circle of the regular hexagon.

Step 4: With a radius equal to the length AB and the compass point at A, mark off point F. Move the compass point to F and mark off point E. Repeat on the other side of the circle to find points C and D.

Step 5: Join B and C, C and D, D and E, E and F and also F and A.

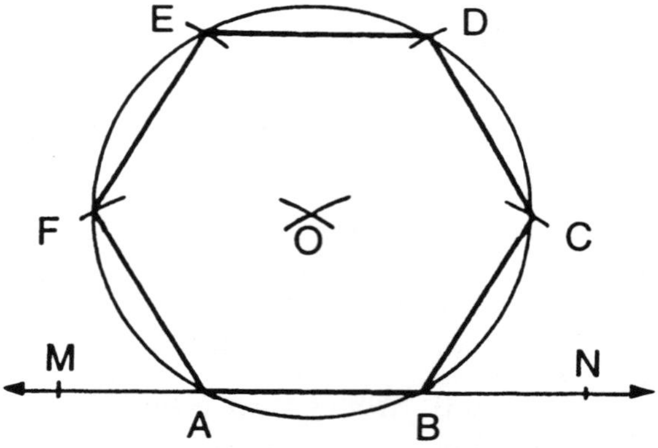

Fig 5.18

2.6 *Octagons (Eight-sided figures)*

How to construct a regular octagon

Step 1: Draw a circle with D as the diameter.

Step 2: Construct horizontal, vertical amd 45° lines through the centre of the circle.

Step 3: Use a 45° set square to draw the sides of the octagon.

The line segments comprising the sides of the octagon must be drawn in such a way that they will just touch the circle. The octagon can also be drawn when a side is given, using the same method as for the hexagon, but with a 45° set square.

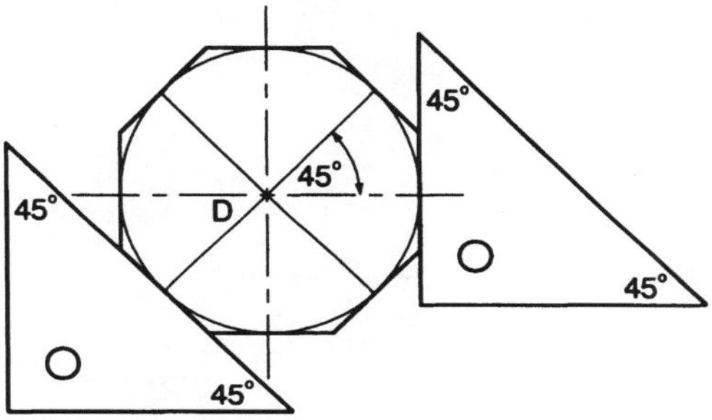

Fig 5.19

2.7 *Joining straight lines with arcs using a compass*

How to construct an arc of a circle so that it touches two lines that form a right angle

Step 1: Draw a right angle. Position the compass point on the point of the angle and draw an arc with the **desired radius**.

Step 2: From each point where the arc intersects with the straight lines, draw another arc. Where the two arcs intersect, the centre for the **desired arc** is formed.

Step 3: Place the compass point on the intersection and draw the radius touching the angle.

Step 4: Join the arc with the straight lines.

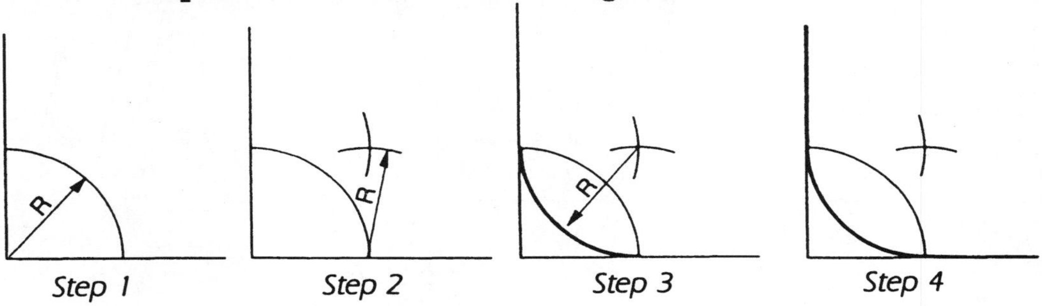

Fig 5.20

How to construct an arc of a circle so that it touches lines that form an acute angle

Step 1: Draw an acute angle. Then draw lines parallel to these lines at a distance equal to the desired radius, R, of the arc.

Step 2: Drop perpendiculars from the intersection point to the lines forming the angle.

Step 3: Draw the required arc from the intersection point.

Step 4: Join the arc with the straight lines of the angle.

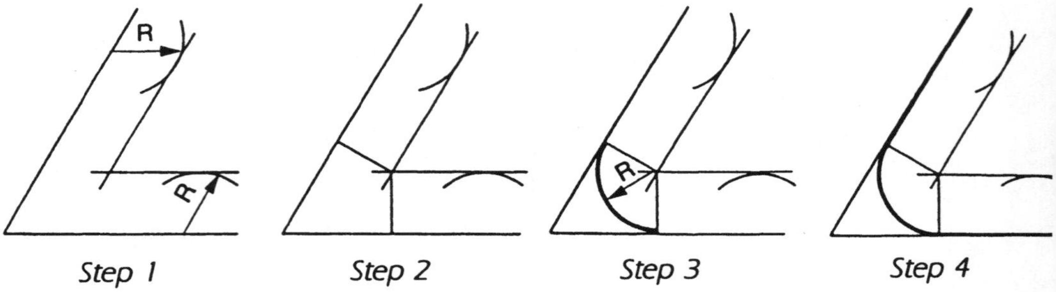

Fig 5.2.1

How to construct an arc of a circle so that it touches lines that form an obtuse angle

Step 1: Draw an obtuse angle. Then draw lines parallel to these lines at a distance equal to the desired radius, R, of the arc.

Step 2: Drop perpendiculars from the intersection point to the lines forming the angle.

Step 3: Draw the required arc from the intersection point.

Step 4: Join the arc with the straight lines of the angle.

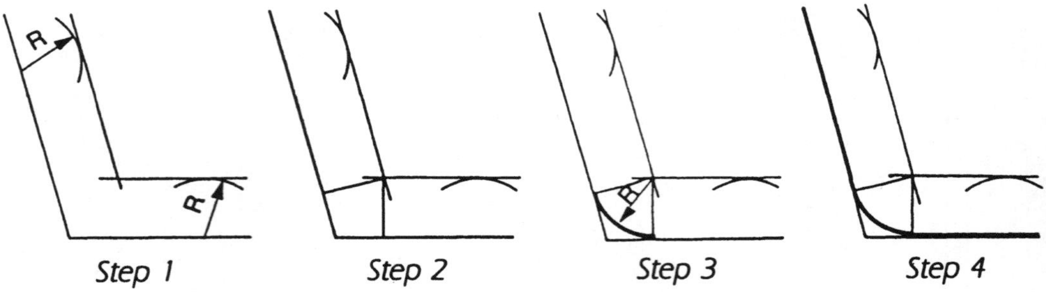

Fig 5.22

Activity 7

Show all the constructions clearly.

1 Construct full scale the drawings shown below.

2 Insert dimensions.

Use only your drawing instruments. Do not use a radii template for drawing arcs. (A radii template is a plastic mould shaped like a radius.)

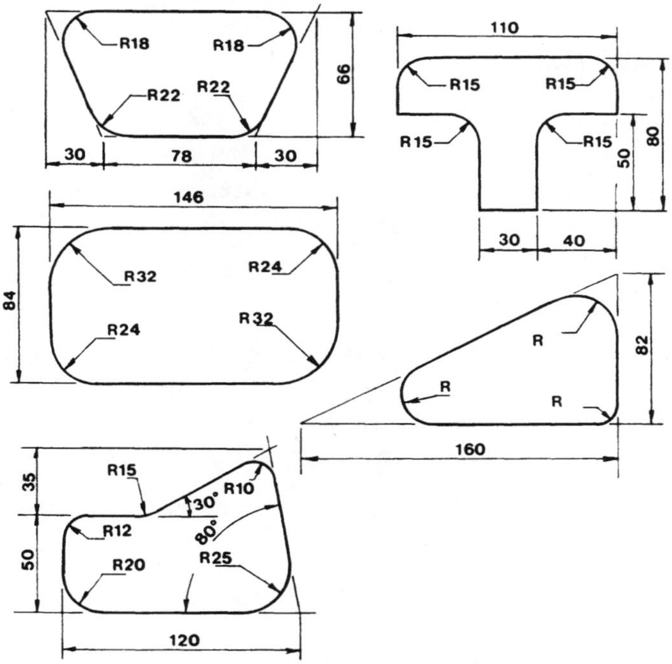

Fig 5.23

How to construct an arc of a circle so that it touches a line and another arc or a circle

Step 1: Draw the arc or circle with radius R1 and the line as construction lines.

Step 2: Draw a construction line parallel to the straight line at a distance equal to the required radius, R. With radius = R + R1, draw an arc of a circle, as shown in fig 5.24. Where the new arc and parallel line meet will be the centre of the desired arc.

Step 3: Draw the arc.

Step 4: Darken the lines.

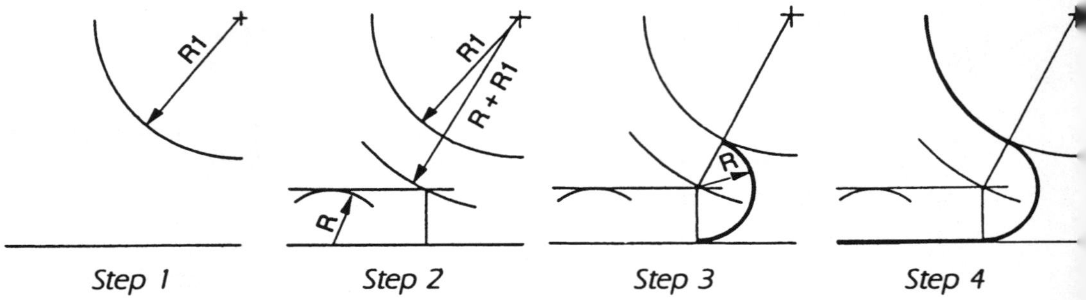

Fig 5.24

How to construct an arc of a circle so that it touches two other arcs or circles

Step 1: Draw the given arcs, or circles, as construction lines, as shown.

Step 2: Add the radius of the required arc of the circle to the radius of the original two arcs and draw two circular arcs to intersect each other. Join this point of intersection with each centre point of the two arcs. All of this is done with construction lines.

Step 3: Draw the required arc of the circle using the intersection point as a centre.

Step 4: Darken the other arcs, or circles, as a full line.

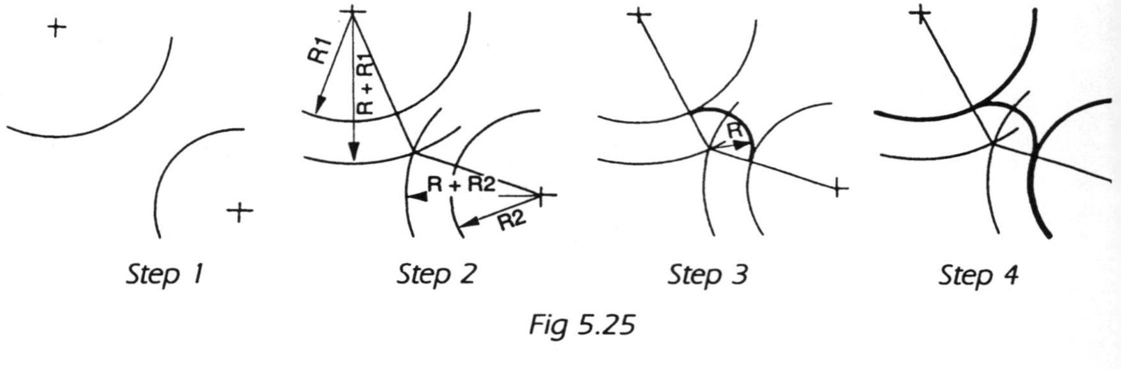

Fig 5.25

Activity 8

Show all the construction lines.

1 Construct full scale the drawings shown in fig 5.26.
2 Insert the dimensions.

Use only your drawing instruments. Do not use a radii template for drawing the arcs.

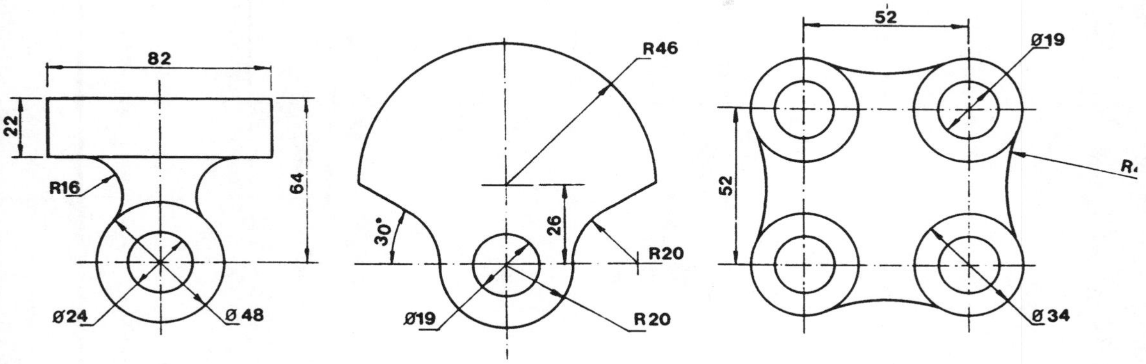

Fig 5.26

2.8 An ellipse

Unlike a **circle** in which the horizontal and vertical axes are equal in length, the two axes of an ellipse are different in length. The long axis is called the **major axis** and the short axis is called the **minor axis**. In this course we are going to use the **trammel** and **four-centre methods** to construct ellipses.

Trammel method

A trammel is an instrument used for drawing ellipses. In this case we are going to use a home-made trammel made out of cardboard.

Step 1: Set out the major axis AB and minor axis CD.

Step 2: Cut one end of a long piece of cardboard into a sharp point and mark off half the major axis to equal **a** and half the minor axis to equal **c**.

Step 3: Now use the trammel across the axes so that one of the marks is always on the major axis and the other on the minor axis. For each position make a light pencil mark at the sharp point. For two of the four quadrants, the trammel will have to be turned over and the marks **a** and **c** transferred from one side to the other.

Step 4: Finally, join the marks with a smooth curve to form the ellipse.

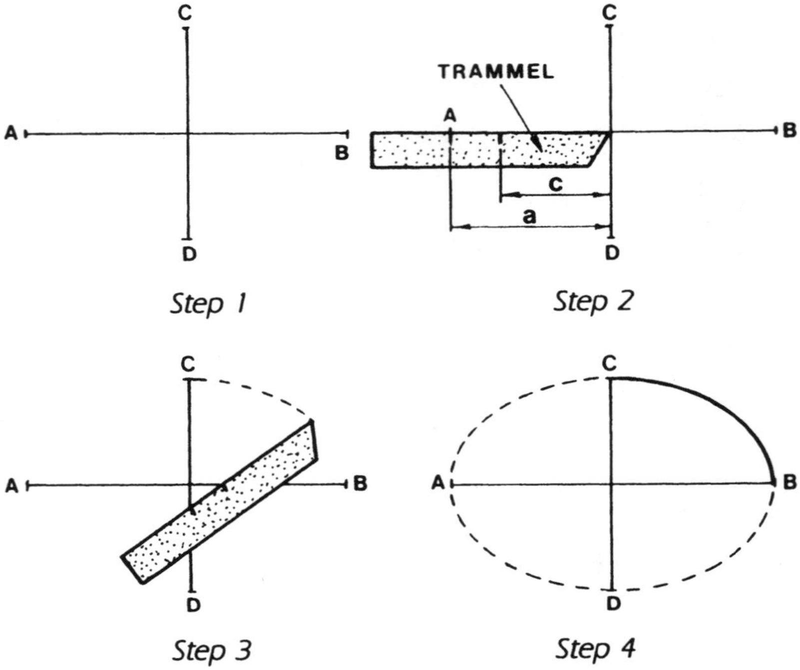

Fig 5.27

Four-centre method

For this method you will need a compass to construct the ellipse.

Step 1: Mark off the major axis AB and minor axis CD.

Step 2: Join point A of the major axis to point C at the end of the minor axis.

Step 3: With point O as centre and OB as radius, draw an arc from point B to obtain point E on OC produced (which means OC has been extended).

Step 4: With point C as centre and CE as radius, draw an arc to obtain point F.

Step 5: Bisect AF to obtain points G and J.

Step 6: Mark off OH, which is equal to OG. Join J to H and extend.

Step 7: Mark off OK, which is equal to OJ. Join K to G and H and extend.

Step 8: With the compass point on J and a radius of JC, draw the top arc. Move the compass point to K and draw the bottom arc.

With the compass point at G and a radius of GA, draw the arc on the left side. Move the compass point to H and draw the arc on the right.

Note that the lines GK, HK, GJ and HJ indicate the points where the arcs join together.

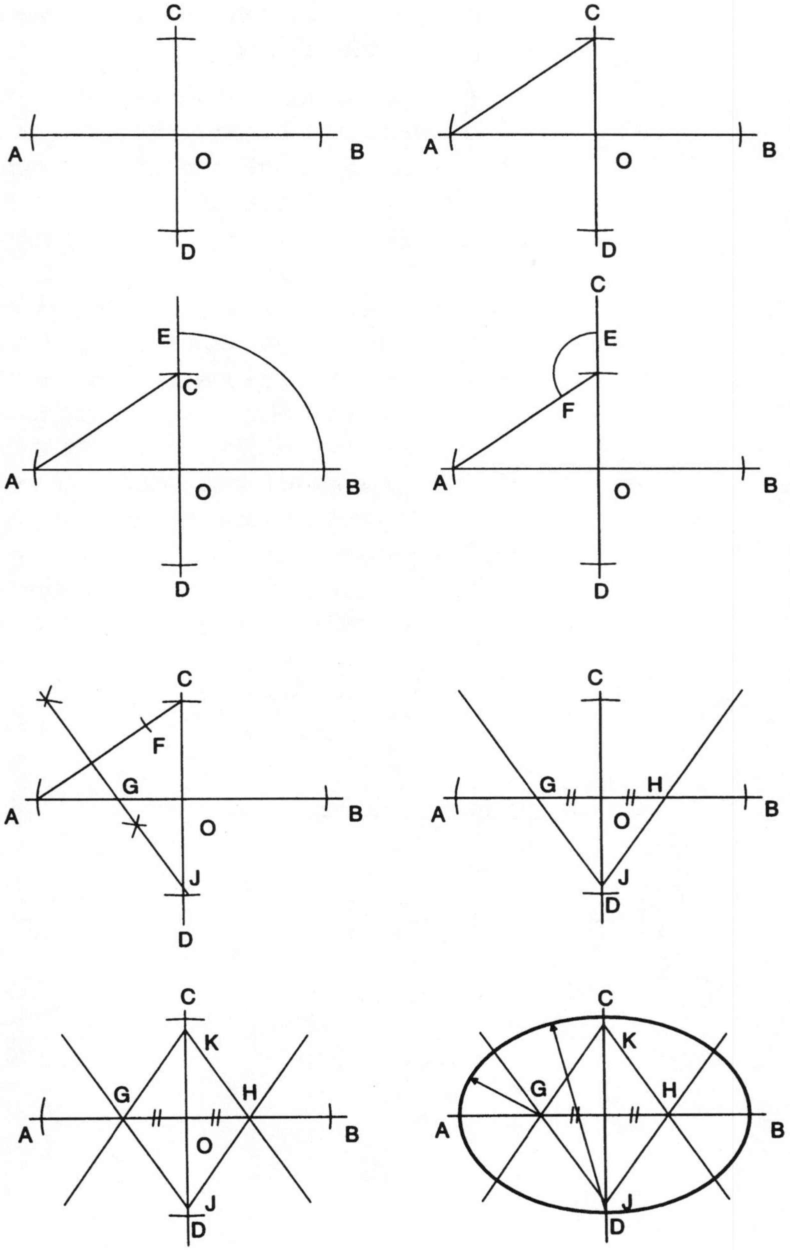

Fig 5.28

Activity 9

As in the case of the other activities, you must practise these constructions until you can draw them almost without thinking. The secret is to practise every second day and not be discouraged if you don't succeed in your first attempts. Remember you will not be assessed on whether you **can list the steps of construction**, but on whether you **can draw an accurate construction**. Draw all the constructions full scale.

1 Use the trammel method to set out an ellipse with a major axis of 140 mm and a minor axis of 85 mm.
2 Use the four-centre method to construct the following ellipse: major axis 100 mm, minor axis 50 mm.
3 Construct an ellipse with a major axis of 100 mm and a minor axis of 75 mm.
4 Construct the view of the elliptical gland, as shown in fig 5.29a. Insert dimensions and print a title.
5 The profile of the special cam shown in fig 5.29b is made up of half an ellipse and a semicircle. Draw the view of the cam shown. Dimension the view and print a title.
6 Fig 5.29c shows a view of a cover. Construct the given view, insert the dimensions and print the title.
7 The pattern shown in fig 5.29d is made up of a circle containing parts of two identical ellipses, with a major axis of 100 mm and minor axis of 50 mm. Draw the pattern.

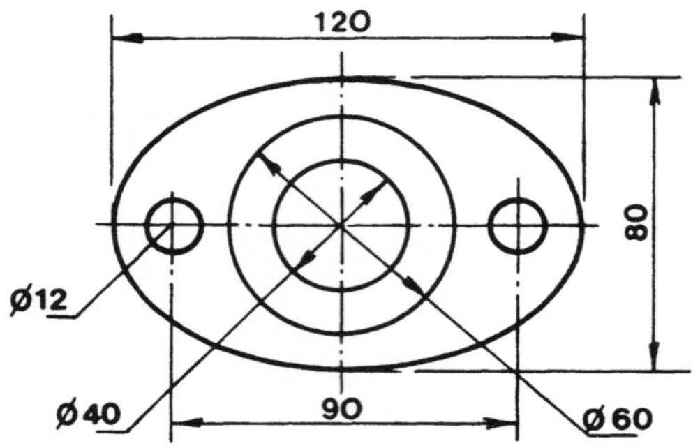

a

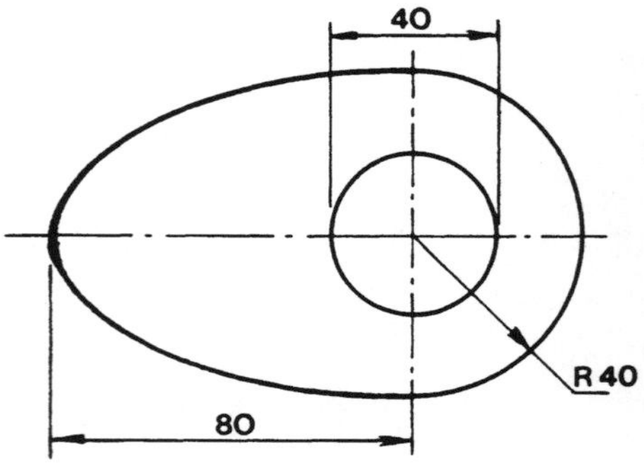

b

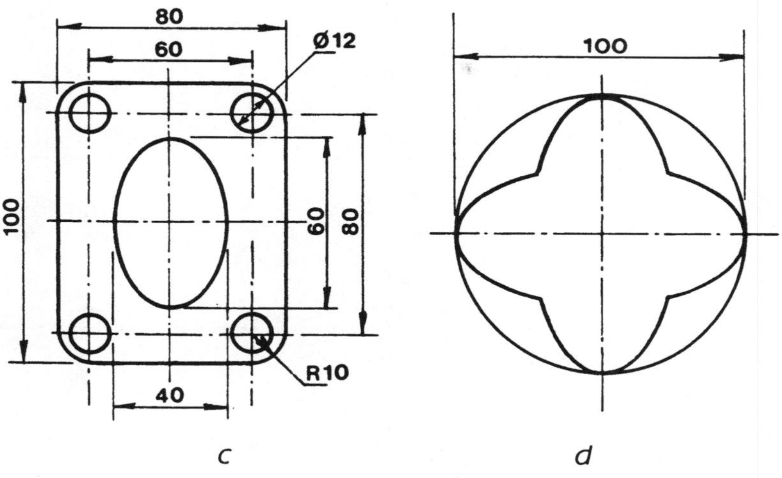

Fig 5.29a–d Views of constructions

3 Summary

Having completed unit 5, which deals with the various kinds of constructions used in drawing, you have been provided with all the basic theories and skills which underpin drawing as a subject. In the remaining units, where you will be asked to construct various working drawings, you will have to apply your newly gained knowledge and techniques.

Self-evaluation

- ◇ Practise the exercises contained in the nine activities until you can do them quite easily.
- ◇ As in the previous units, you should draw up and create your own review sheets of the unit for quick reference. By going over these sheets on a weekly basis to start off with, and thereafter on a monthly basis, you will find it much easier to cope with the examination load at the end of the year.
- ◇ Review the material presented in fig 5.30 and be sure that you are familiar with all the terms.
- ◇ If you find it difficult to understand this unit, you must contact someone who can help you. You can consult other students, your lecturer, technicians or draughtspersons.

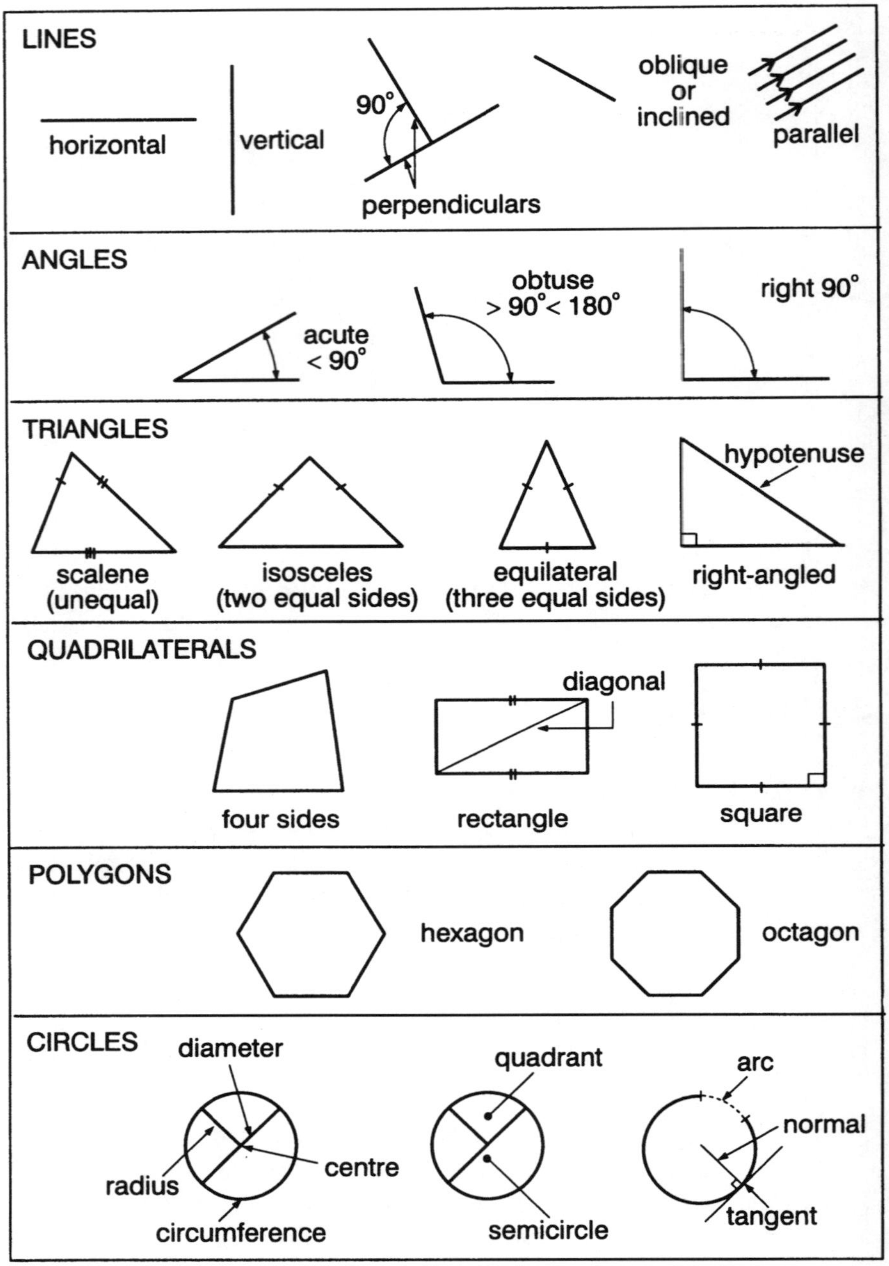
LINES
horizontal
vertical
90°
perpendiculars
oblique
or
inclined
parallel
ANGLES
acute
< 90°
obtuse
> 90°< 180°
right 90°
TRIANGLES
scalene
(unequal)
isosceles
(two equal sides)
equilateral
(three equal sides)
hypotenuse
right-angled
QUADRILATERALS
four sides
diagonal
rectangle
square
POLYGONS
hexagon
octagon
CIRCLES
diameter
radius
centre
circumference
quadrant
semicircle
arc
normal
tangent

Fig 5.30

Unit 6

Layout of drawings

Study objectives

After studying this unit, you should be able to:

- ◇ Apply first-angle orthographic projections
- ◇ Apply third-angle orthographic projections
- ◇ Lay out a working drawing
- ◇ Draw isometric drawings
- ◇ Draw oblique drawings
- ◇ Project prisms and pyramids
- ◇ Develop pipes
- ◇ Construct interpenetration lines.

1 Introduction

This study unit introduces you to the different types of layout for drawings. You cannot do a proper drawing if you do not know the different layouts. In unit 4, where we discussed freehand drawings, we already started to deal with the different layouts. **Pictorial drawings** such as oblique, perspective and isometric drawings are like **photographs**. They show objects as they would appear to the eye of the observer (see fig 6.1 a–c).

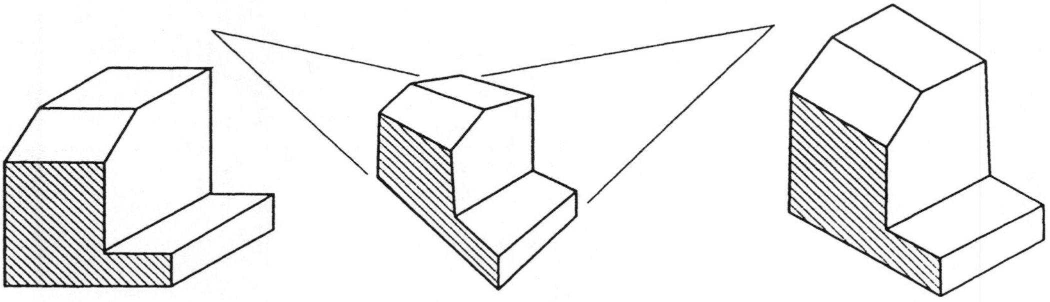

a Oblique *b Perspective* *c Isometric*

Fig 6.1a–c

Such drawings are not often used for engineering drawings, since complicated details are difficult to understand and the dimensioning is difficult to apply. To solve these problems we use orthographic drawings.

What are orthographic drawings?

Orthographic drawings are used to show the **exact shape** of objects. This usually cannot be accomplished in just one view because many details of the object may be **hidden** or **not clearly shown** when the object is viewed from only one side. The **draughtsperson** must therefore show a number of views of the object, seen from different directions. These views are referred to as **front view**, **top view** and **side view** (left or right) and they are systematically arranged on the drawing sheet and projected from one another. These types of projections are called **orthographic projections** or **drawings**.

The standard symbol for indicating the type of projection is shown in fig 6.2. The dimensions are included for your information but should not appear on the drawing.

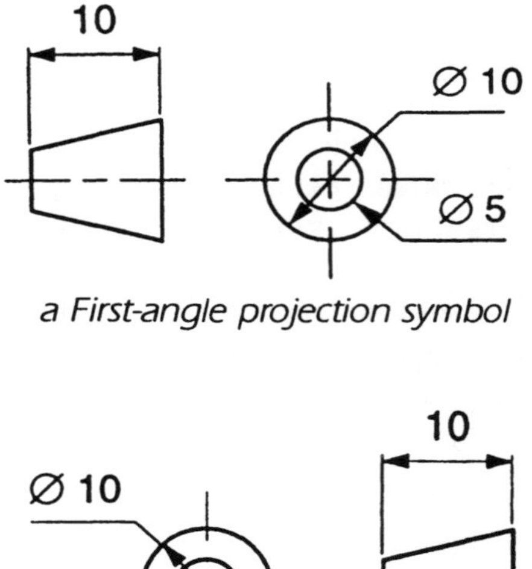

a First-angle projection symbol

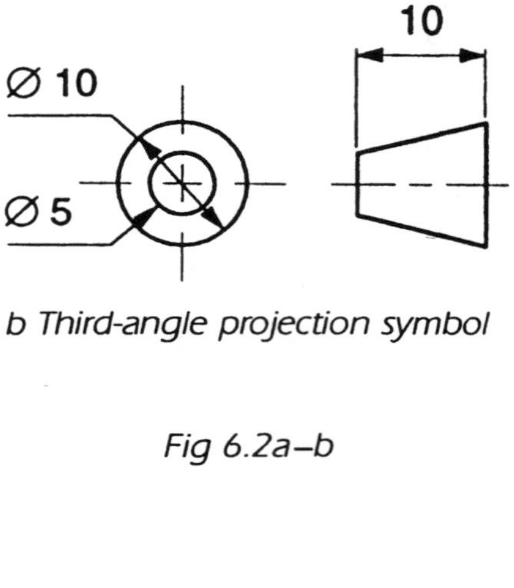

b Third-angle projection symbol

Fig 6.2a–b

2 Layout of drawings

2.1 First-angle orthographic projection

Fig 6.3 shows a isometric view of an object. To draw this object orthographically, you need to show the true shape of the object. This is done by looking **perpendicularly** at the different sides. To achieve this, you must look in the direction of the arrows. The arrows are perpendicular (at right angles) to the faces or planes.

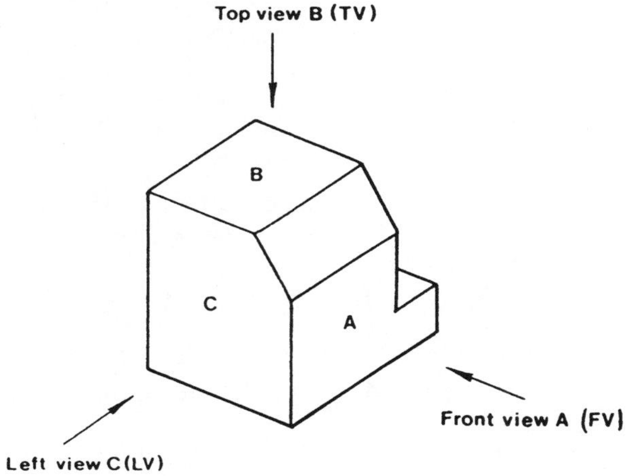

Fig 6.3 Isometric view

To show this object in first-angle orthographic projection, apply the steps below:

Step 1: Take the front view as reference.

Step 2: Place the left view on the right-hand side of the front view.

Step 3: Place the top view below the front view.

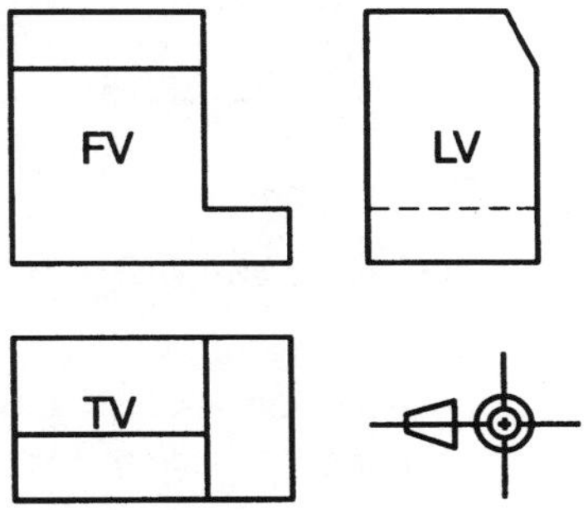

Fig 6.4 First-angle orthographic projection

2.2 Third-angle orthographic projection

A pictorial representation of the different views is shown in fig 6.5

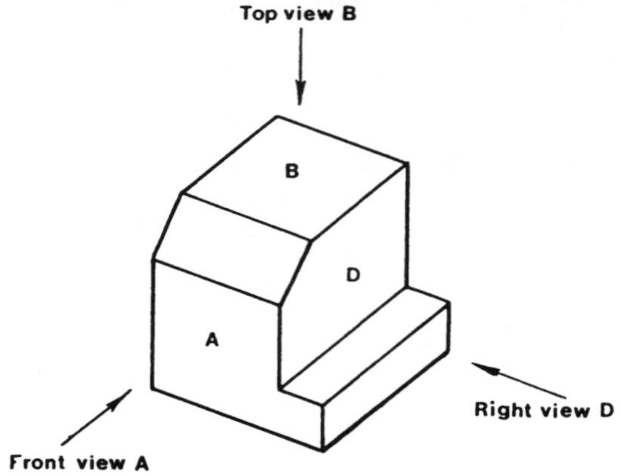

Fig 6.5 The arrows perpendicular (at right angles) to the faces or planes

To show fig 6.5 in third-angle orthographic projection, apply the steps below:

Step 1: Again, take the front view as a reference.

Step 2: Place the right view on the right-hand side of the front view.

Step 3: Place the top view at the top of the front view.

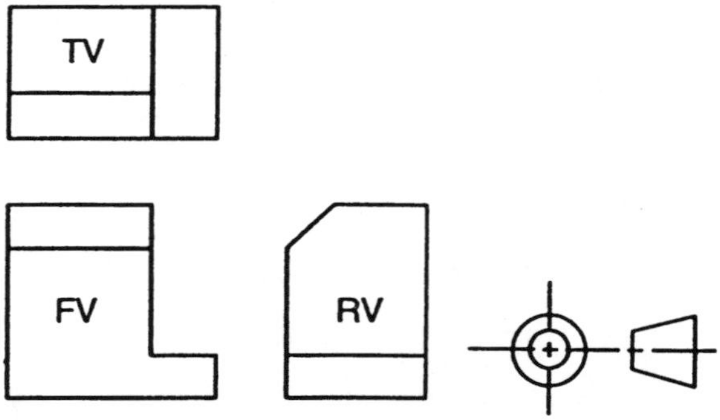

Fig 6.6

The reason for the different arrangements of views is the following:

- ◇ First-angle orthorgraphic is projected in the first quadrant
- ◇ Third-angle orthographic is projected in the third quadrant

Example 1 Draw the object shown in the isometric drawing (fig 6.7a) in first-angle orthographic projection. Include a front view, and a left view and a top view.

Each side of the triangle on the isometric grid, as shown, is equal to 5 mm.

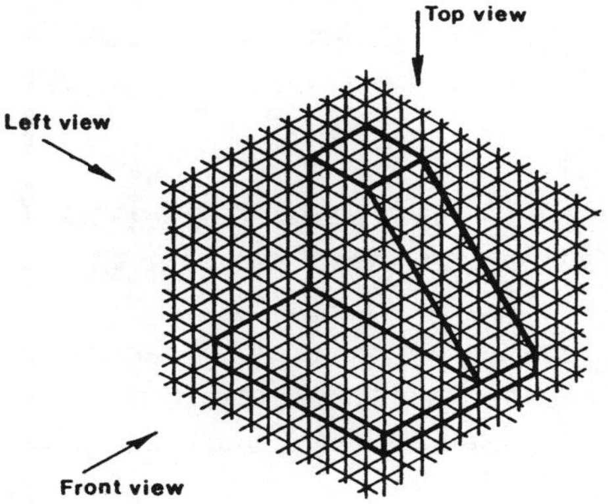

a Isometric view

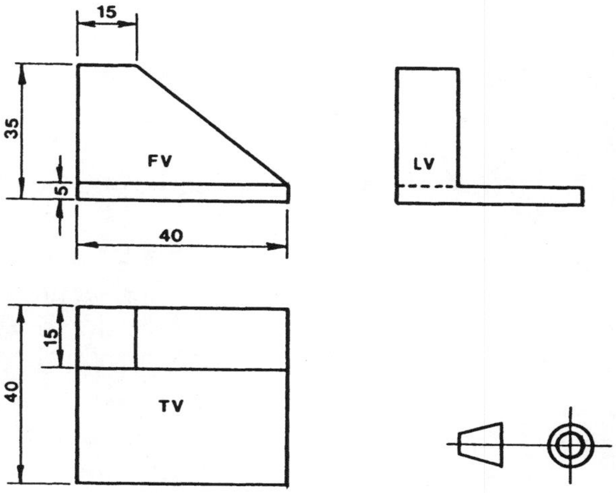

b First-angle orthographic projection

Fig 6.7a–b

Example 2 Draw the object shown in fig 6.8a in third-angle orthographic projection. Include the front view, right view and top view.

Each side of the table on the isometric grid, as shown, is equal to 5 mm.

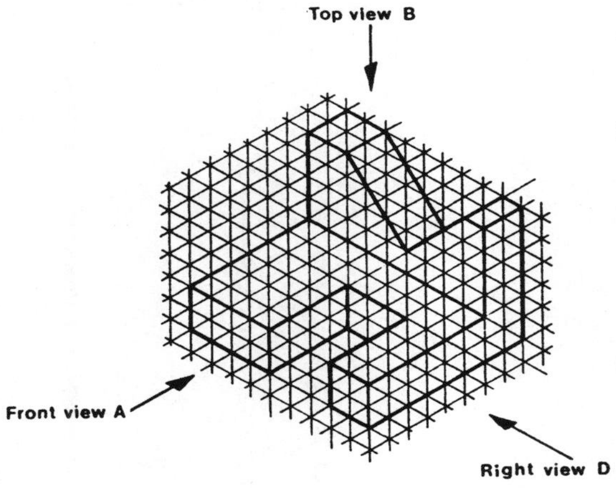

a Isometric view

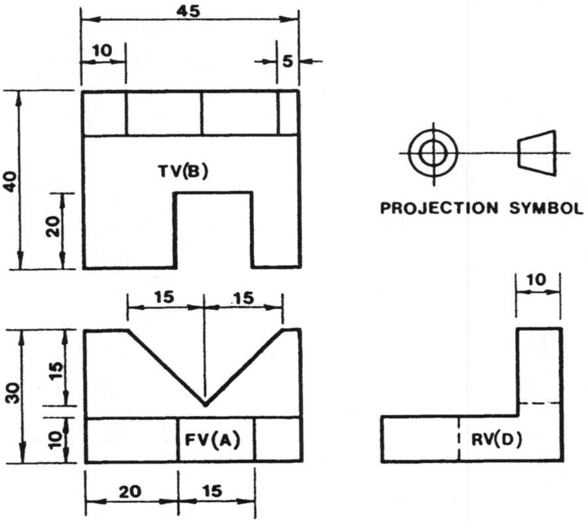

b Third-angle orthographic projection

Fig 6.8a–b

The above examples are only a guide to help you to complete the activity below. The only difference is that you are going to work with real dimensions using all your instruments and will not use a grid, as shown in the examples. We used the grid to make it easier for you to understand the steps involved in drawing first- and third-angle projections.

Activity 1

In this project you will be drawing orthographic projections. Do not copy the isometric drawing, start directly with orthographic views. In all cases, the front view will be in the direction of arrow A. Plan the layout carefully.

1. Fig 6.9 shows a figure in **isometric projection**. Draw in first-angle projection the three views of the object. Insert the dimensions. Scale 1:1.
2. Fig 6.10 shows an isometric view of a bracket. Draw in a first-angle orthographic projection the front view, left view and top view of the bracket. Insert the title and scale, as well as the dimensions. Also insert the projection symbol. Scale = full scale.
3. Fig 6.11 shows a figure in isometric projection. In a third-angle orthographic projection, show the three principle views. Insert the dimensions. Scale 1:1.
4. Fig 6.12 shows a isometric presentation of a bearing block. Draw in third-angle orthographic projection the front view, top view and right view of the bearing block. Insert the title, the scale and the symbol in the correct place. Also insert the dimensions. Scale 1:1.

Note: not all hidden lines are shown

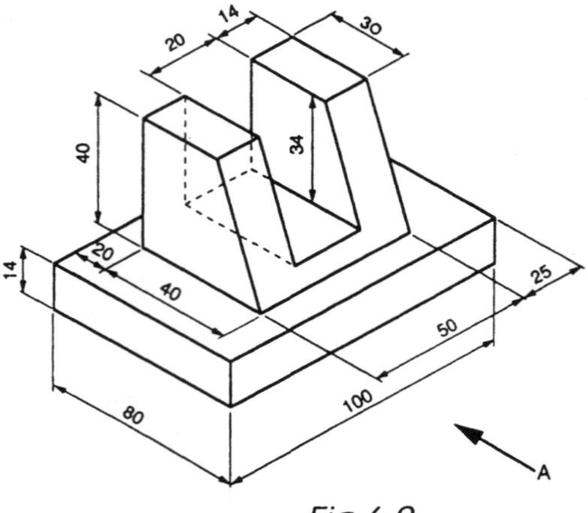

Fig 6.9

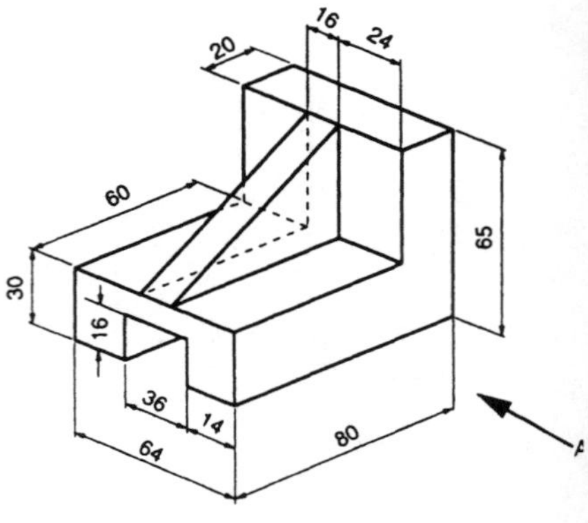

Fig 6.10

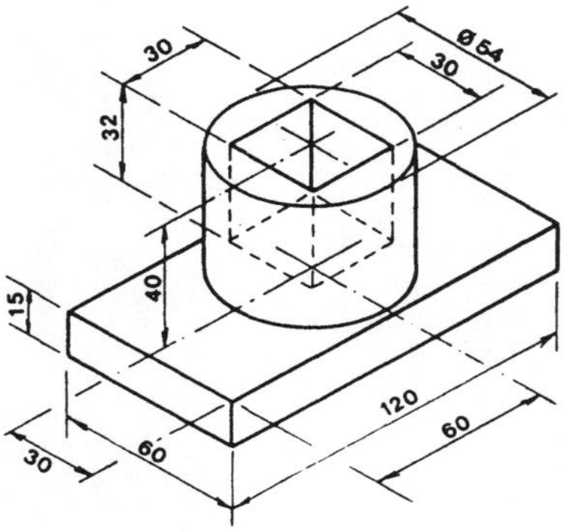

Fig 6.11

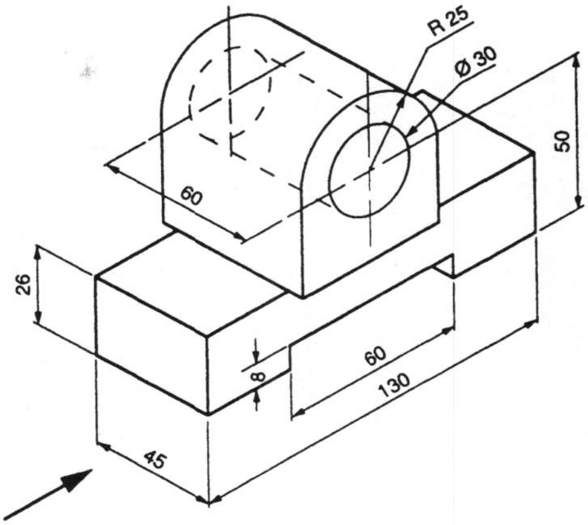

Fig 6.12

2.3 Projecting a third view

Sometimes two views of an object are given and you are required to find the third view. To do this you should make use of the projection lines to project the third view. The **projection lines** are **faint** so that they do not interfere with the outlines of the object. For instance in fig 6.13 the front and top views are given in first-angle orthographic projection. Find the third view.

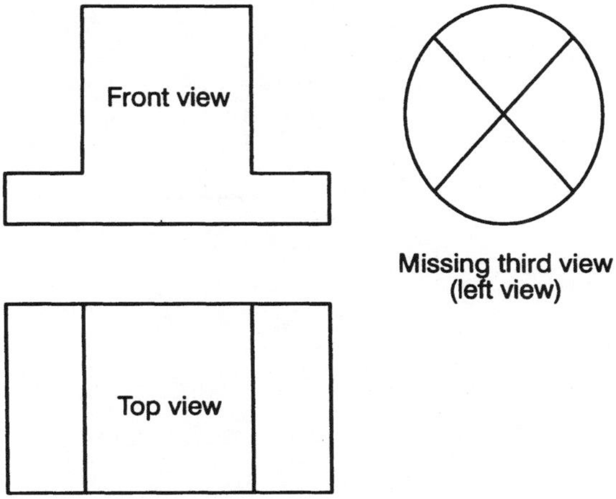

Fig 6.13

Now, with the aid of construction projection lines and a 45° mitre line, you will be able to find the third view. Look at fig 6.14, which explains how to find such a view. Refer to unit 2 section 2.7 to review the use of projection lines and mitre lines.

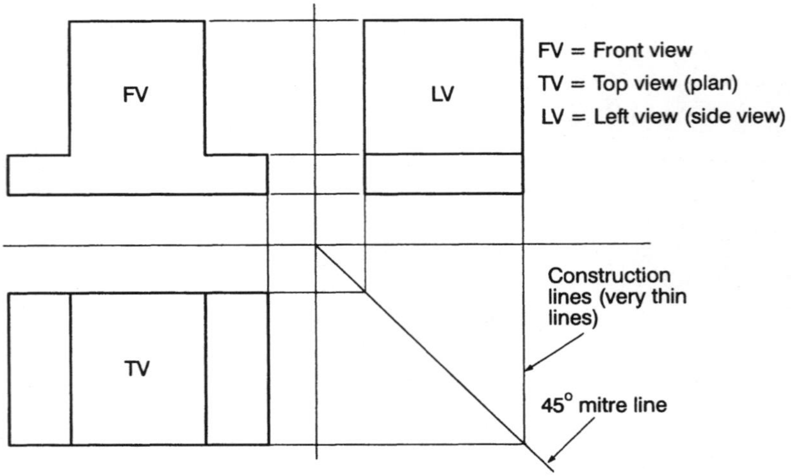

Fig 6.14

Activity 2

Two views of an object are given in first-angle orthographic projection. Construct the two given views and find the third view, as explained in fig 6.14. Scale 1:1.

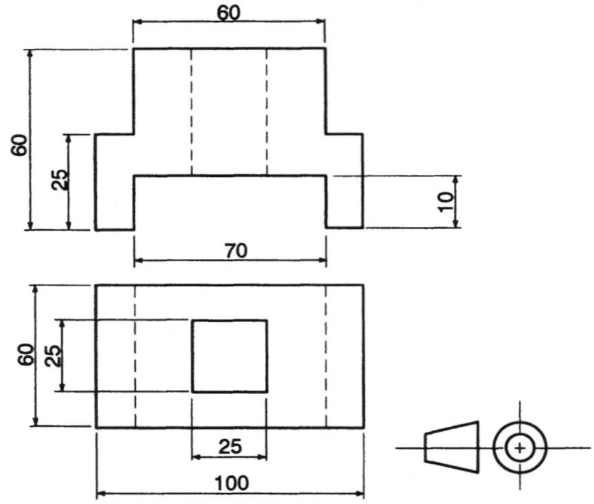

Fig 6.15

2.4 Drawing step-method for laying out drawings

The drawing step-method provides a logical sequence for ensuring a quality end result.

Fig 6.16 shows a condensed version of the drawing step-method.

Step 1: Start with the **centre lines** and the block layout. Measure from the centre lines outwards. Complete the layout in very thin construction lines. Insert circles and radii.

Step 2: Erase the unnecessary lines and draw the outlines.

Step 3: Insert the **dimensions**.

Step 3: Insert the **dimensions**.

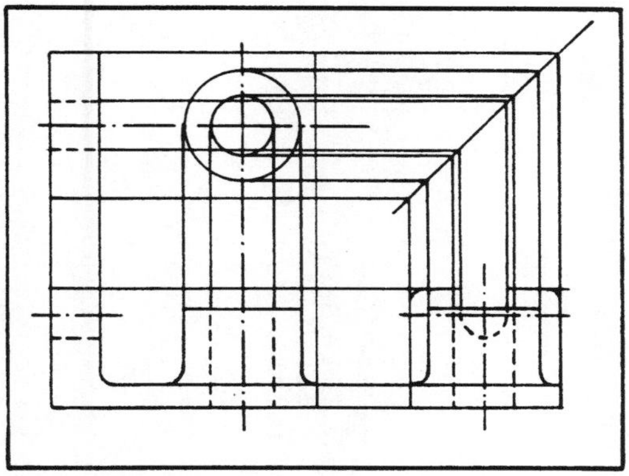

Step 1

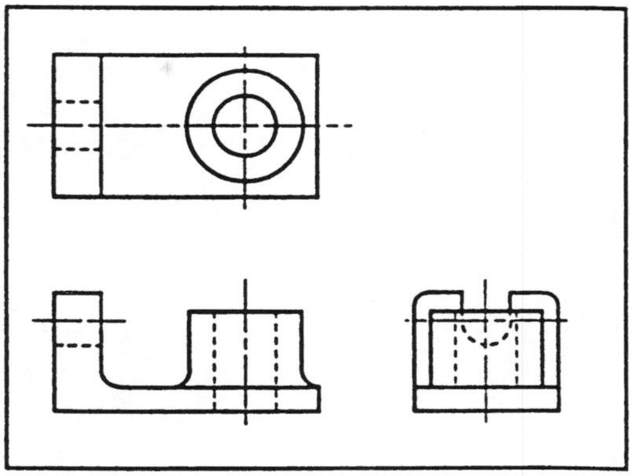

Step 2

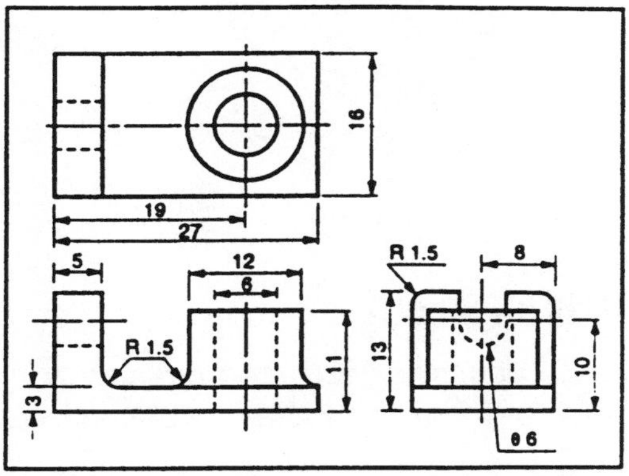

Step 3

Fig 6.16

2.5 Isometric drawings

Although **orthographic drawings** are standard practice for working drawings, **isometric drawings** are much more easily understood by those not trained to read orthographic drawings. The advantage of isometric drawings is that they show **all three planes or faces** in one single view. In fig 6.16, two views are shown in first-angle orthographic projection. Steps 1-3 show you how to draw these views isometrically.

Step 1: Start with the isometric 'box' that will contain the object.

Step 2: Set out important dimensions.

Step 3: Draw in lines and darken final outlines.

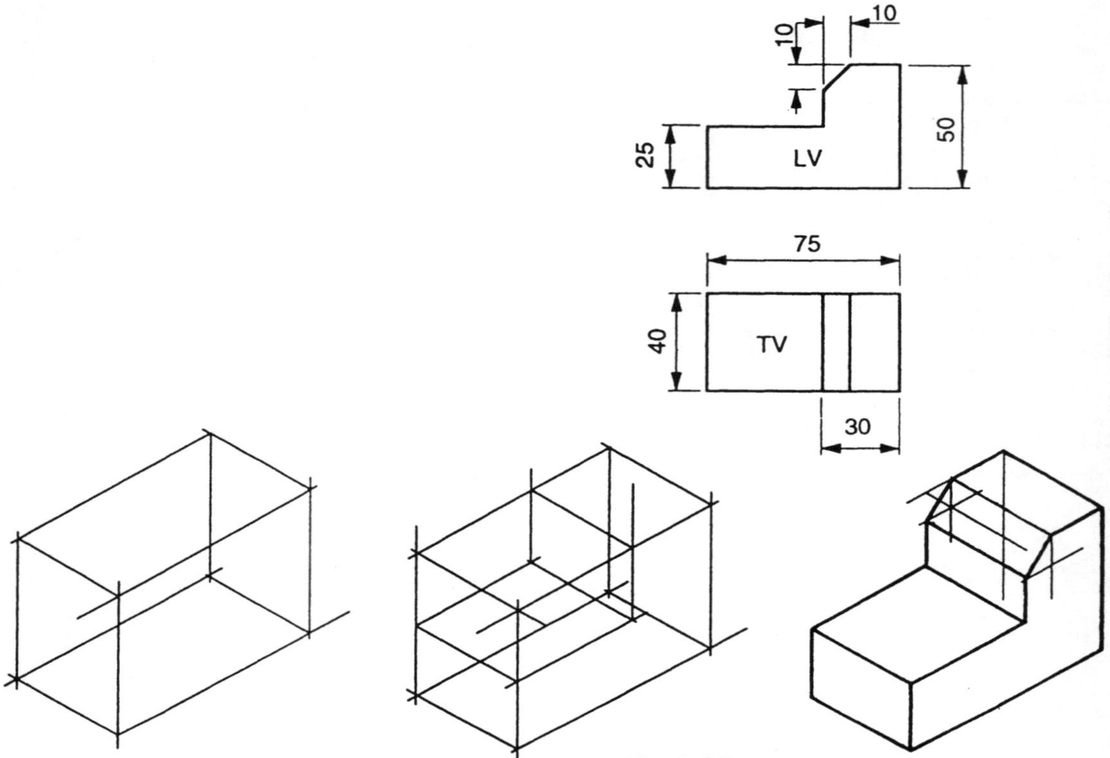

Fig 6.17

Isometric circles

Circles will appear as **ellipses** in isometric drawings. The reason for this is that you are looking at the circle from an angle.

An isometric view of any object shows three planes on one drawing. If you look at fig 6.18, you will see isometric circles on all three planes.

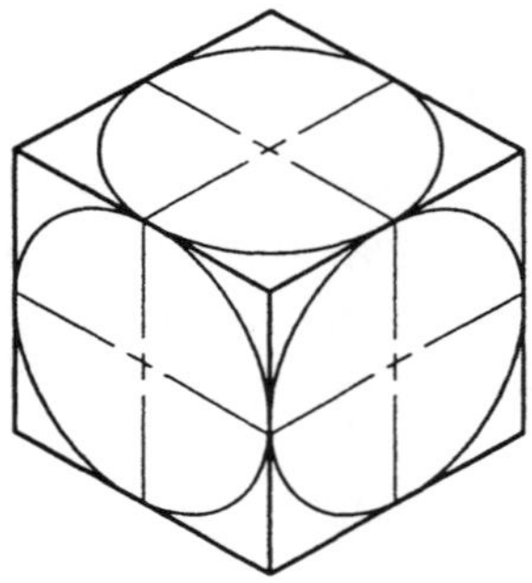

Fig 6.18

To construct these circles on all three planes, follow the **four steps** below. Fig 6.19 shows an orthographic view of a circle, 50 mm in diameter, enclosed in a square.

Step 1: Draw the centre lines and enclosing square in isometric view.

Step 2: Join the obtuse corners to the midpoints of the opposite sides.

Step 3: Two of the four centres will be found where these lines intersect. Do you remember how we discussed the 4-centre method for drawing an ellipse in unit 5 (see section 2.8). Now, the two other centres are the obtuse corners themselves.

Step 4: Complete the approximate ellipse with the four centres obtained.

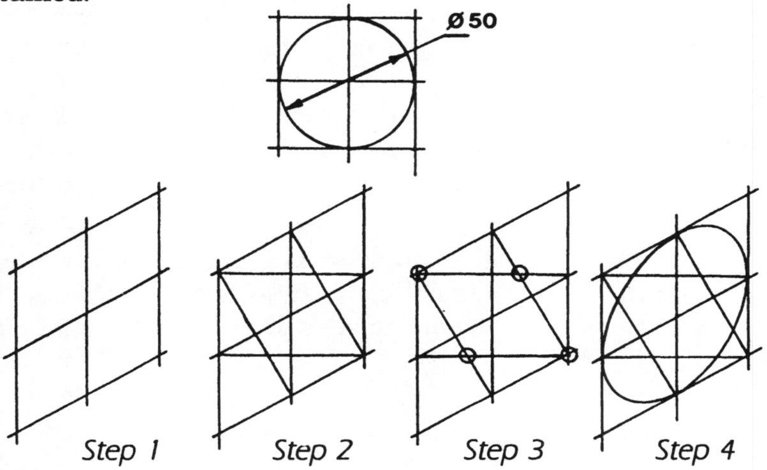

Fig 6.19

Follow the above steps for the other two planes (see fig 6.20).

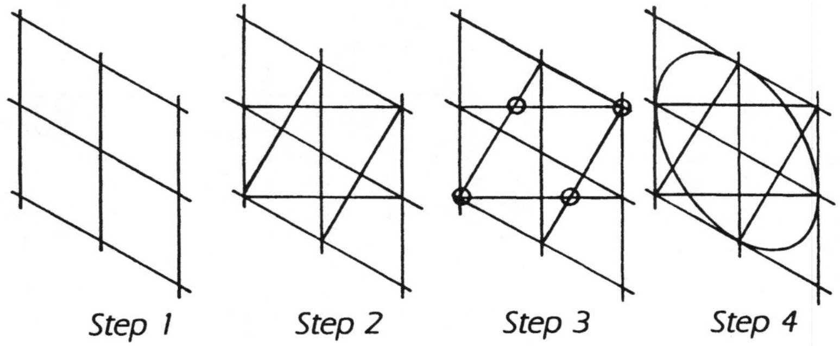

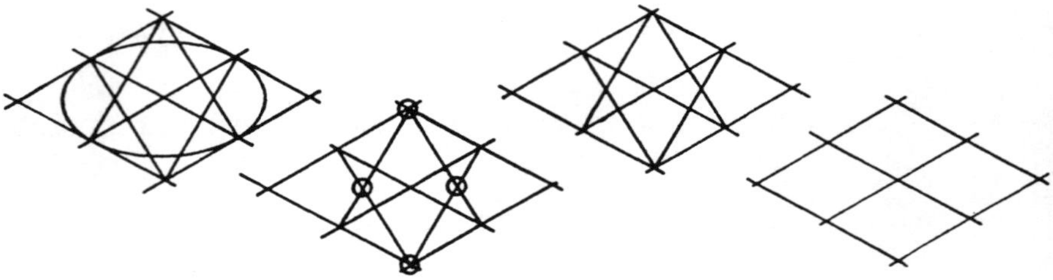

Fig 6.20

Activity 3

1 The object in fig 6.21 is shown in third-angle orthographic projection. Follow the steps discussed above to draw this object isometrically. The scale used will be full scale. Do not insert dimensions and do not copy the orthographic views. Show only the final drawing.

2 The front view and left view of an object are given in first-angle orthographic projection, as shown in fig 6.21. Do not copy these two views, but simply draw the final isometric view. Do not insert dimensions. Scale 1:1.

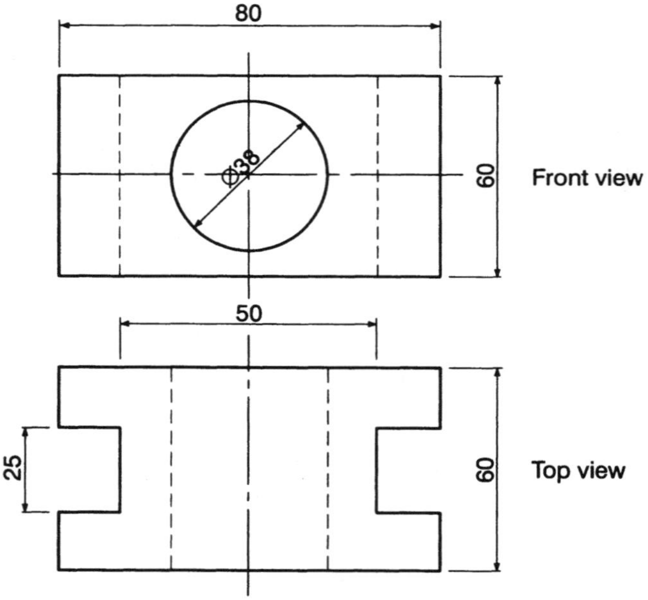

Fig 6.21

3 The front views and left views are shown in fig 6.22. They are drawn in first-angle orthographic projection. Do not copy the views given. Just draw the isometric drawing. Do not insert dimensions. Scale 1:1.

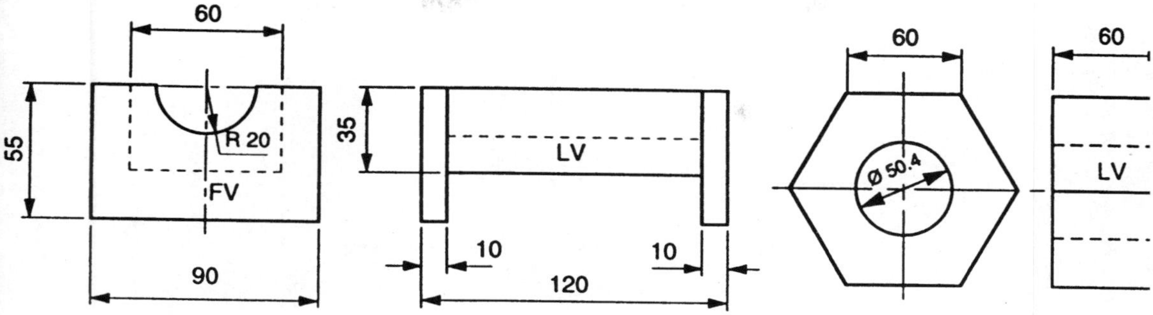

Fig 6.22

2.6 Oblique drawings

In isometric drawings, lines are shown **vertically** and at **30° to the horizontal**. In oblique drawings, the lines are drawn **vertically, horizontally and at 45° to the horizontal**. Another difference between isometric and oblique is that in oblique drawings the measurements along the 45° lines are made half the true length (refer to unit 4, section 3.7). In figs 6.23 and 6.24 the differences are clearly indicated.

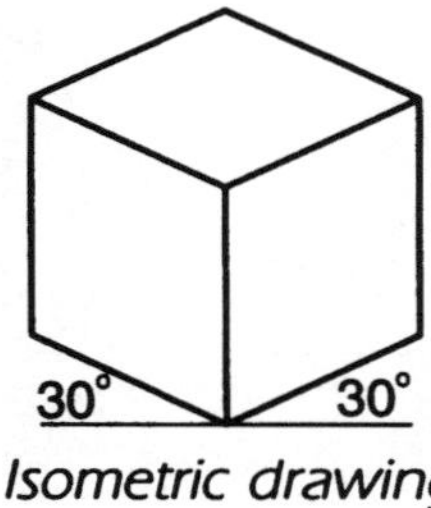

Isometric drawing

Fig 6.23

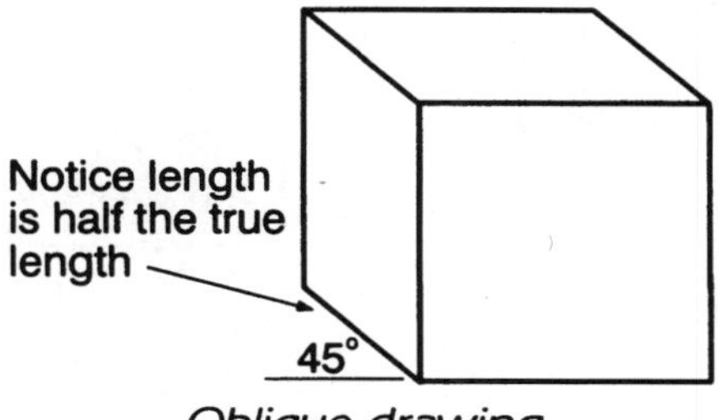

Oblique drawing

Fig 6.24

Example 3 An object is shown in fig 6.25. Two views are given in first-angle orthographic projection. Make an oblique drawing. Scale 1:1.

Step 1: Start with the oblique 'box' that will contain the object. One side is at 45°.

Step 2: Along the edges of the 'box' mark off the dimensions given. Remember, halve the dimensions on the 45° line.

Step 3: Draw the outlines and indicate hidden edges.

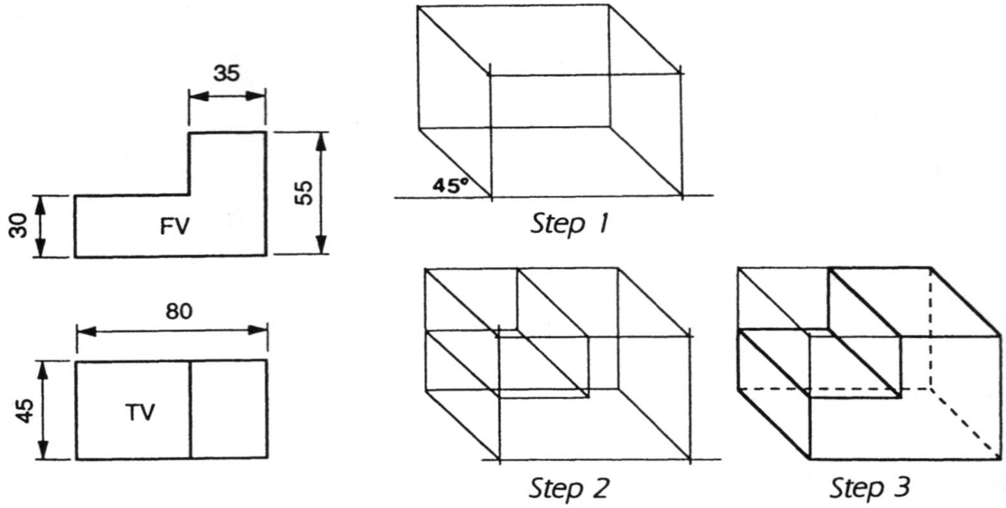

Fig 6.25

Example 4 An object is shown in first-angle orthographic projection in fig 6.26. Follow the steps above to draw an oblique drawing.

Oblique circles on the front view are ordinary circles. Circles on the side and top views are ellipses and are drawn using the four-centre method.

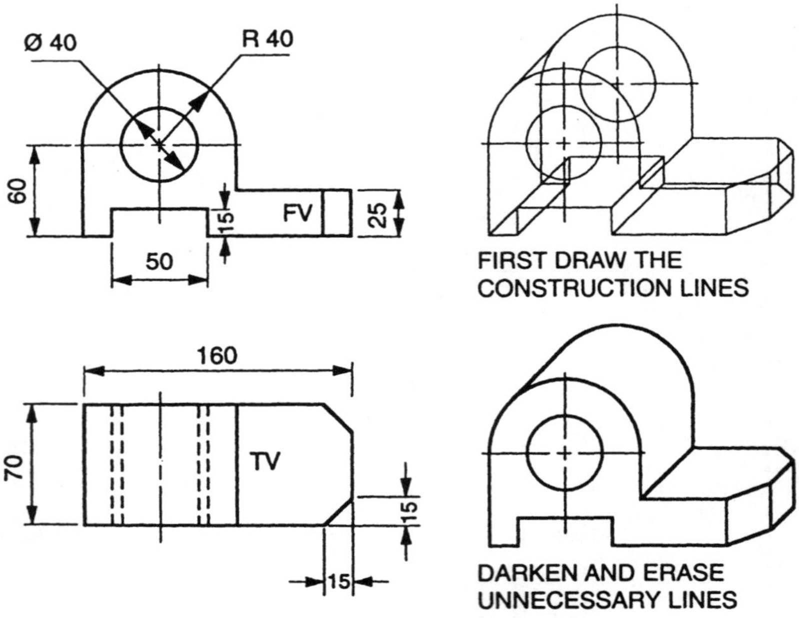

Fig 6.26

Activity 4

Figs 6.27 and 6.28 show two views of an object in first-angle orthographic projection. In each case, draw an oblique drawing. Scale 1:1. Do not insert dimensions and do not copy the orthographic views.

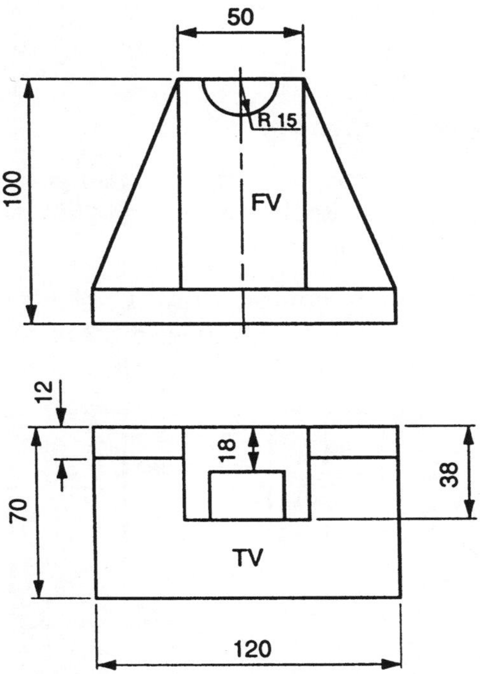

Fig 6.27

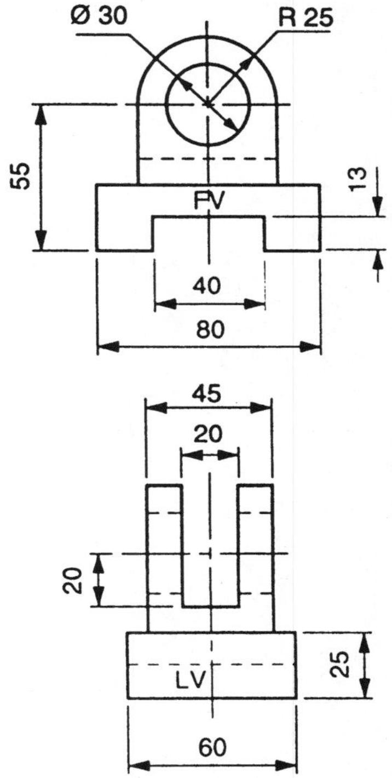

Fig 6.28

2.7 Projections of prisms and pyramids

◇ A **prism** is a solid having **two flat ends** of the same shape. In other words, the top and the bottom of a prism look exactly the same, whether the prism is triangular or octagonal, etc.

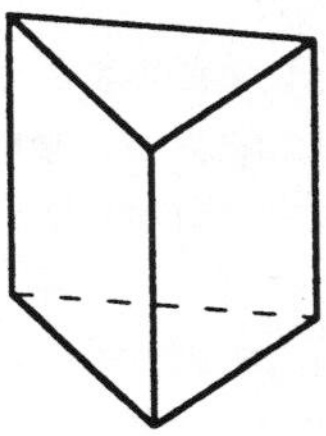

Triangular

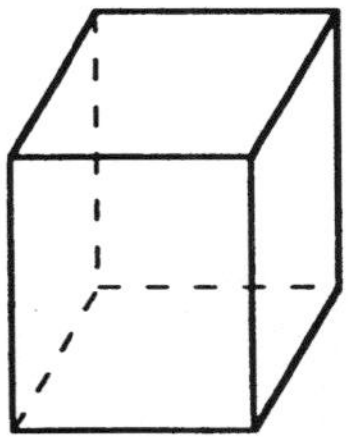

Rectangular

◇ A **pyramid** is a solid having a flat base of any shape and a sharp point called the apex or **vertex**. Pyramids can also be shaped as triangles or octagonals, etc.

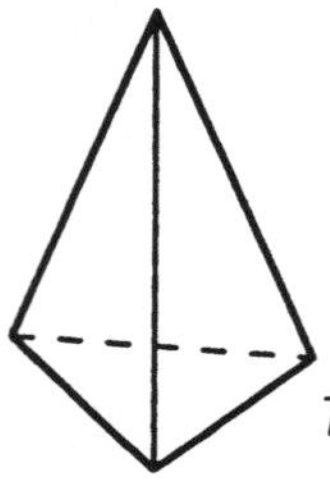

Triangular

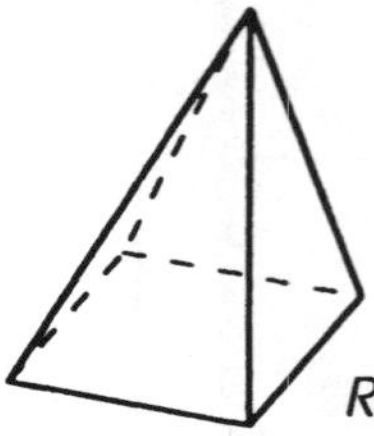

Rectangular

In fig 6.29 you are required to draw a triangular prism. The dimensions will be given. Start with the top view, then do the front view and, finally, the left view.

Example 5

Fig 6.29 shows an orthographic drawing of a triangular prism. The dimensions are shown. You should start with the top view then project the front view and finally the left view.

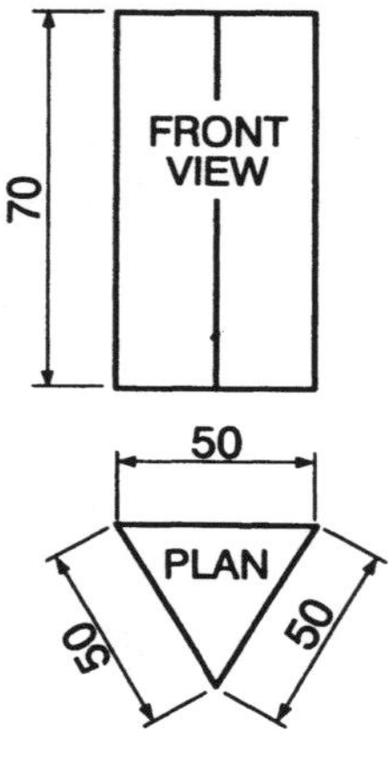

Fig 6.29

Activity 5

1. Fig 6.30 shows a top view and an incomplete front view of a triangular prism. Construct the given top view, complete the front view and project a left view. Dimension the views.
2. Fig 6.31 shows a top view and an incomplete front view of a hexagonal prism. Constuct the given top view, complete the front view and project a left view. Dimension the views.
3. Fig 6.32 shows a top view and an incomplete front view of an octagonal pyramid. Construct the top view, complete the front view and find the left view.

4 Fig 6.33 shows a top view and an incomplete front view of a square pyramid. Construct the top view, complete the front view and project a left view.

All drawings should be drawn full scale.

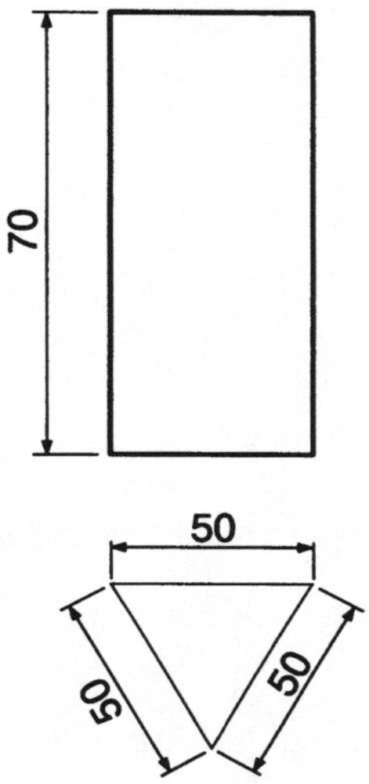

Fig 6.30

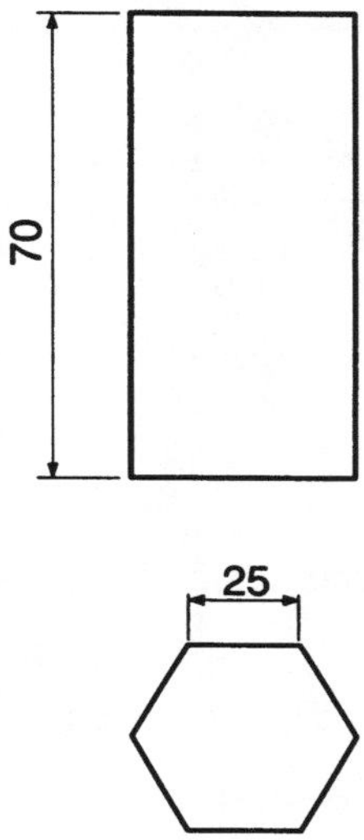

Fig 6.31

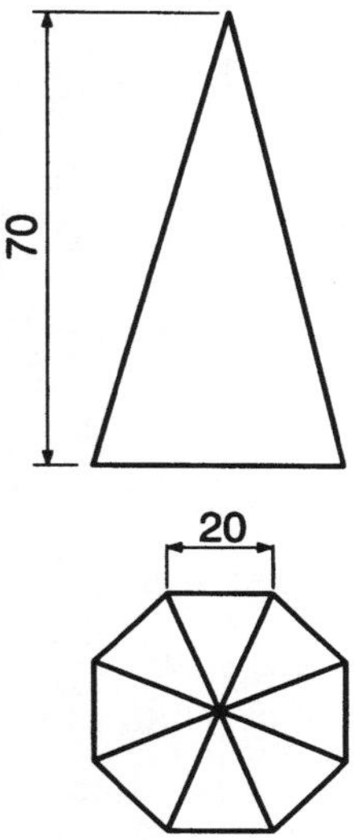

Fig 6.32

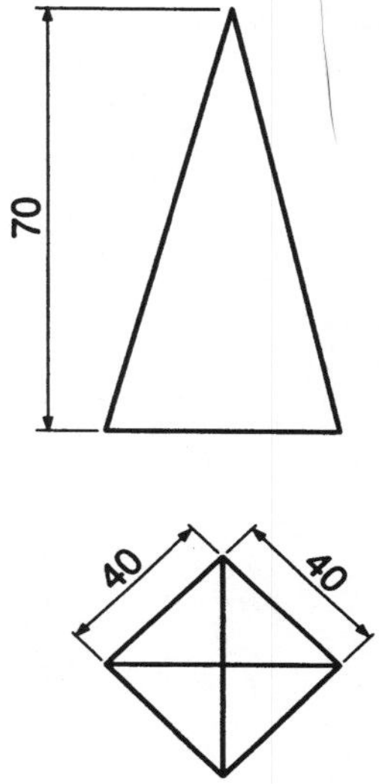

Fig 6.33

2.8 Developments

A 'development' in draughting refers to the layout of a pattern on a sheet-metal plate. If you want to develop a pipe from a sheet-metal plate, you first set out the pattern, cut it to size and form the pipe.

Development of a triangular pipe

To form a triangular pipe, first set out the pattern on a sheet-metal plate, cut out the profile and bend it to the shape of a triangular pipe. Fig 6.34 shows the development using the following steps.

Step 1: Construct the top view first, as you did in activity 5, and finish the front view.

Step 2: Number the top view as shown. Transfer the numbers to the front view.

Step 3 Extend the base and set out the numbers 1-2-3-1 at the true length.

Step 4: Complete the pattern as shown.

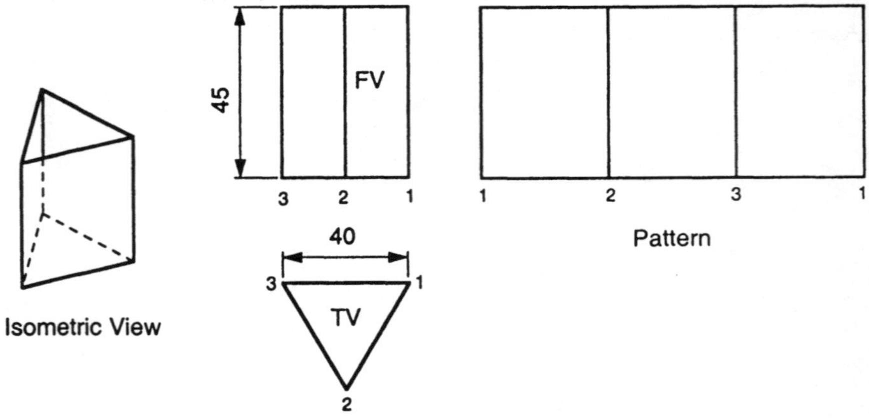

Fig 6.34

Development of a truncated open-ended rectangular pipe

Rectangular means the object is shaped like a square, but one pair of sides is longer than the other.

Truncated means the top part of an object is cut of at an angle.

Open ended means the ends are not closed. You can see through one end to the other.

Two views of a rectangular sheet-metal pipe, cut off at an angle and open at each end, are shown in fig 6.35.

The procedure for the development of a truncated rectangular pipe is the same as for a triangular pipe.

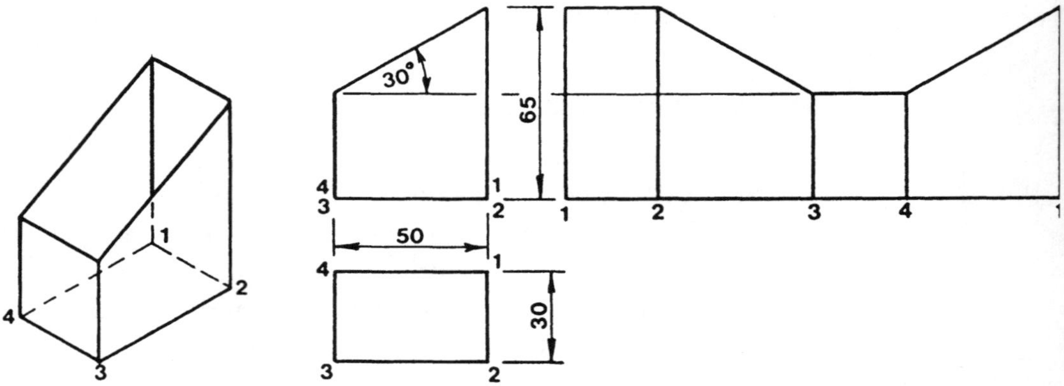

Fig 6.35

Activity 6

Two views of a sheet-metal pipe are given in each case in fig 6.36–6.39. Construct the views and develop the pattern. Remember to start with the top view and then to finish the front view. The scale is 1:1 or full scale.

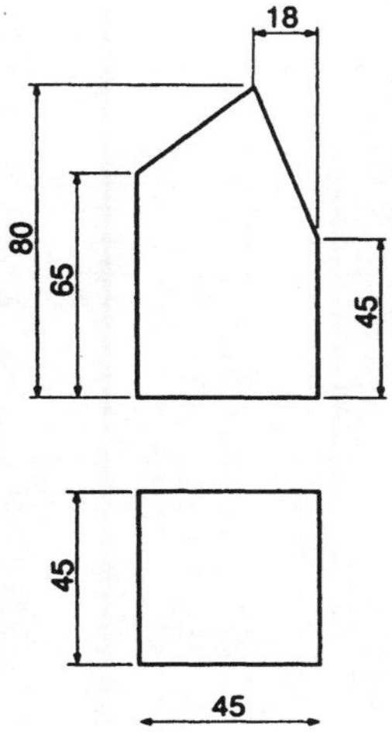

Fig 6.36

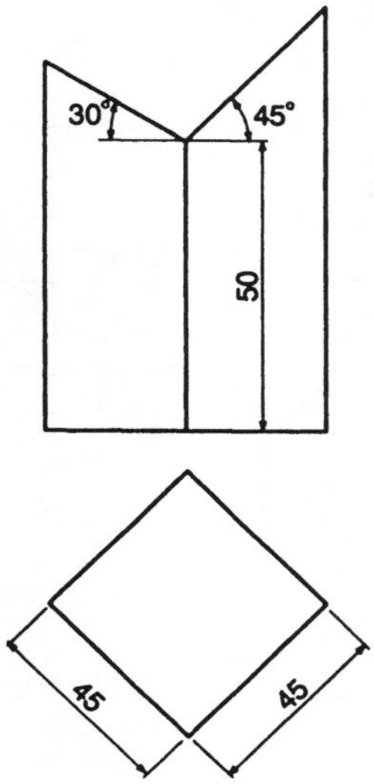

Fig 6.37

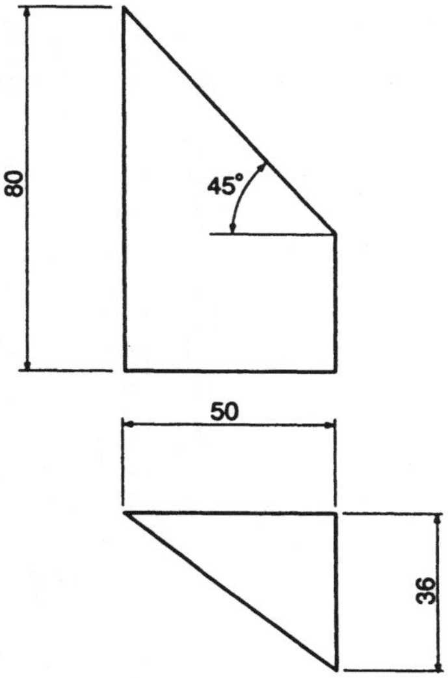

Fig 6.38

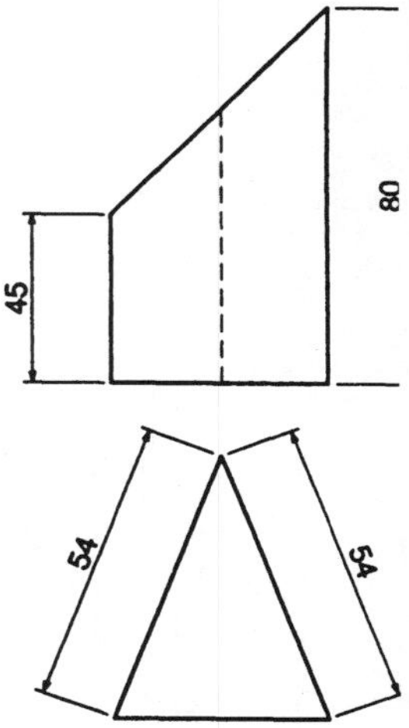

Fig 6.39

Development of a truncated cylinder

Two views of a cylindrical sheet-metal pipe, cut off at an angle and open at both ends, are shown in fig 6.35. The development is show in fig 6.40.

Unlike prisms, cylinders have no corners. You therefore have to draw a number of surface lines on the perimeter of the cylinder which will serve as imaginary edges.

Step 1: Draw the front view and top view to scale. Divide the circumference of the circular top view into 12 equal parts and number the parts as shown in fig 6.40.

From each of the numbered points on the circumference of the circle, project vertically upwards and draw surface lines on the vertical curved surface of the front elevation, as shown.

Step 2: Begin the development by extending the baseline of the front view. Along this line, mark off the circumference of the base of the cylinder by plotting distances 1–2, 2–3, 3–4, . . ., 10–11, 11–12, 12–1 taken from the top view.

Step 3: At points 1, 2, 3 . . ., 11 and 12, draw vertical lines whose true lengths are obtained by projecting across from the corresponding surface lines in the front view. The points of intersection are joined by a smooth curve.

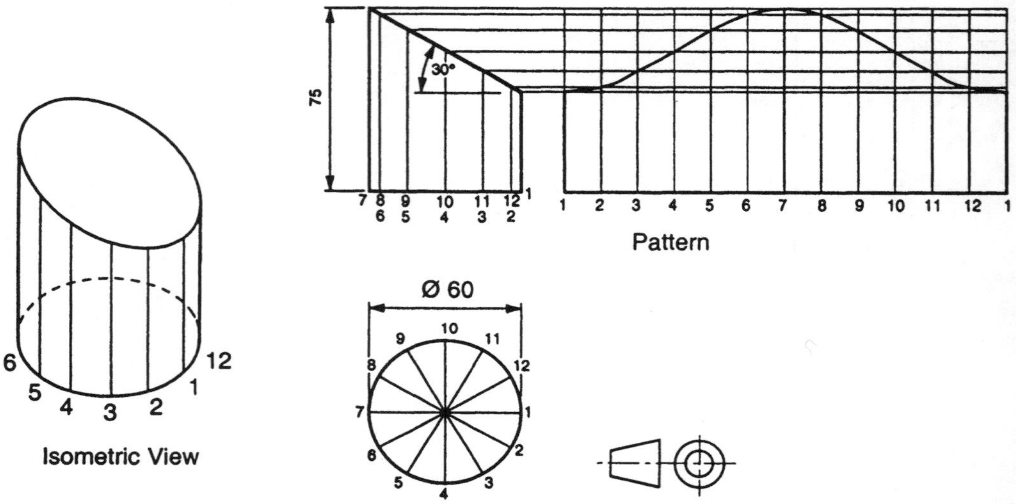

Fig 6.40

Development of a two-piece pipe elbow

Fig 6.41a shows a front view of a cylindrical pipe elbow. The lower piece may be developed in the same way as described for the truncated cylinder in fig 6.41b. The upper piece has both ends cut off at an angle. For convenience, the projection (fig 6.41c) is arranged so that its surface lines are parallel to the axis of the piece.

These types of elbows are used in ducting for air-conditioning and pipes in chemical plants.

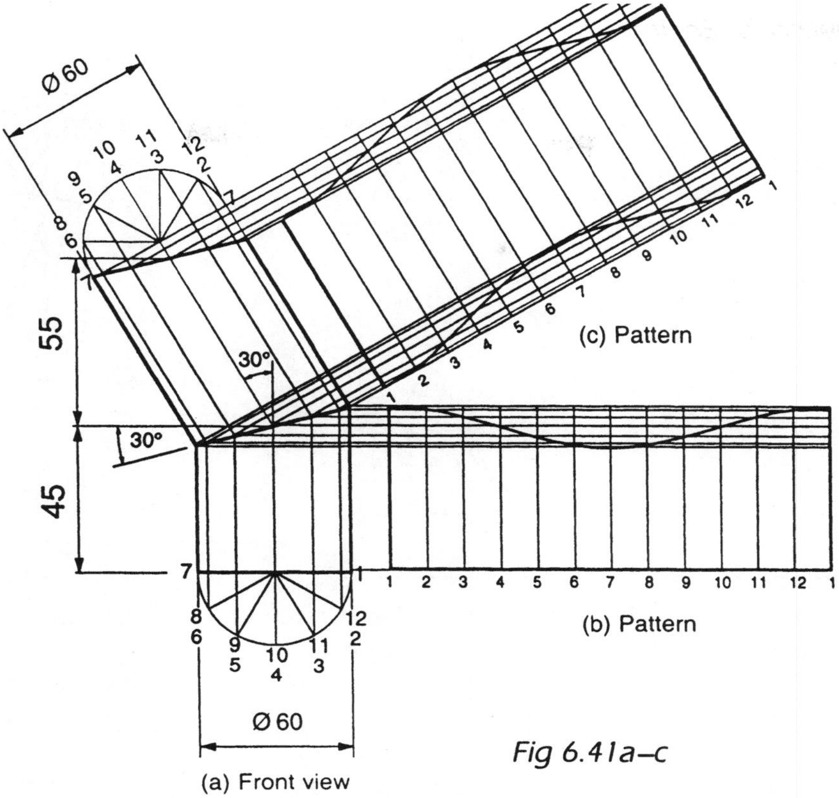

Fig 6.41a–c

Development of a truncated cone

Two views of a truncated cone cut off straight across are shown in fig 6.45a. The development of a curved surface is shown in fig 6.45b.

What type of orthographic projection is used? Look at the projection symbol.

Step 1: Draw the front view and top view. Divide the circumference of the circular base into 12 equal parts and number these points and the apex, as shown in fig 6.42a. From each of these numbered points in the top view, project vertically downwards onto the base of the front view and then draw a radial surface line to meet at the apex O.

Step 2: Begin the development (fig 6.42b) by drawing a circular arc with a radius equal to the slant height O–1 of the cone. Plot distances 1–2, 2–3, 3–4, . . ., 10–11, 11–12, 12–1 taken from the top view onto the arc.

Step 3: The sector O-1-1 obtained in this way is the development of the curved surface of the full cone. The required pattern for the frustum is obtained by removing the smaller sector O-A-A.

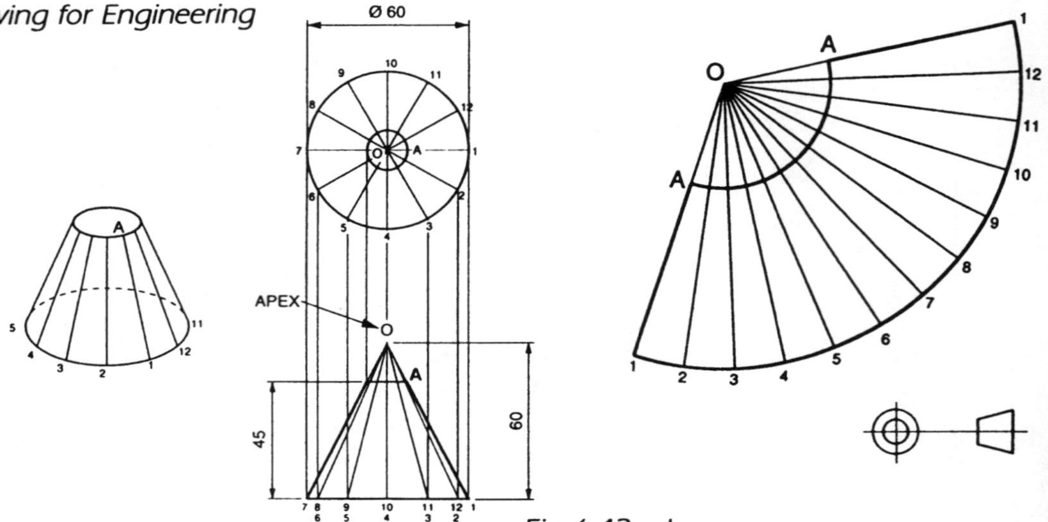

Fig 6.42a–b

Activity 7

1 Develop the surface of the cylindrical sheet-metal duct shown in fig 6.43.
2 Fig 6.44 shows one view of a right-angled pipe elbow. Draw the given view and develop the surface of both pieces.
3 Two parallel horizontal pipes are connected, as shown in fig 6.45. Obtain the patterns for the three pieces.

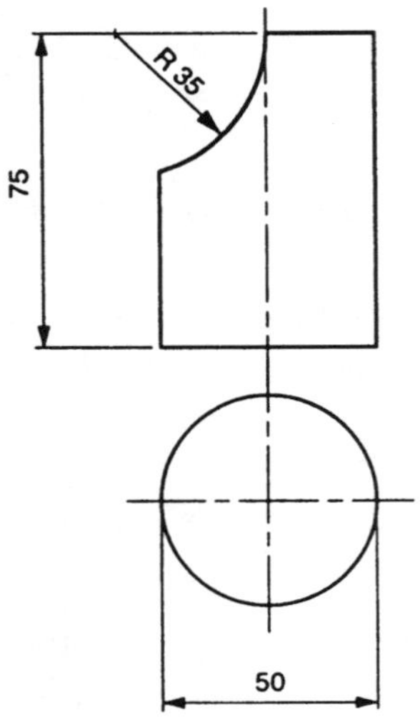

Fig 6.43

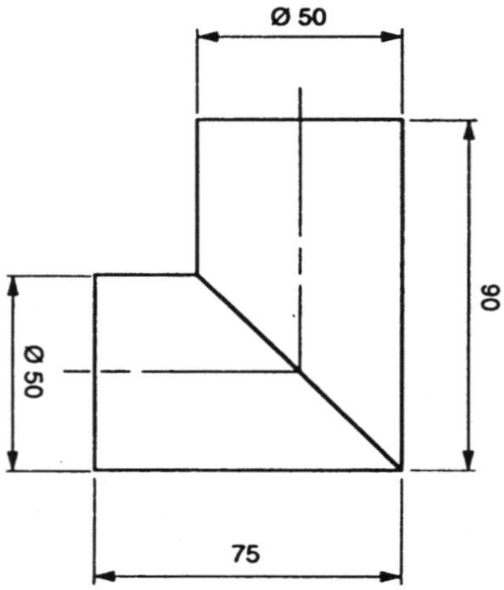

Fig 6.44

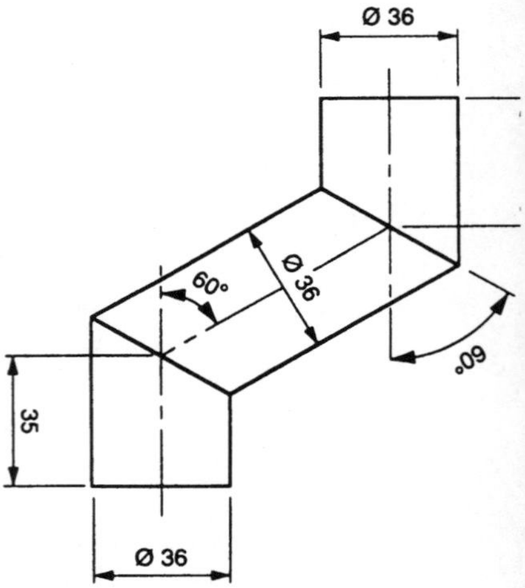

Fig 6.45

4 Copy the two views of the cone shown in fig 6.46 and develop the curved surface.
5 Develop the six sloping sides of the truncated hexagonal pyramid shown in fig 6.47.

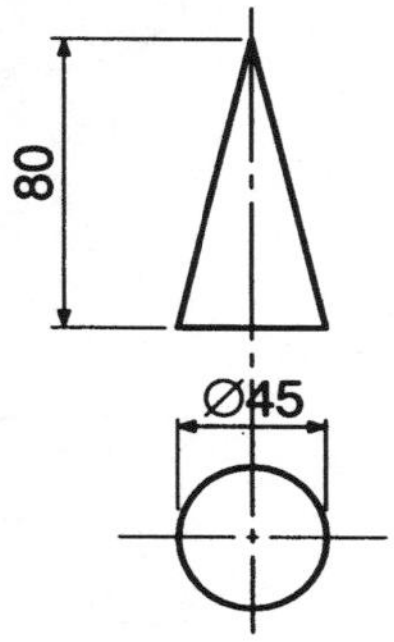

Fig 6.46

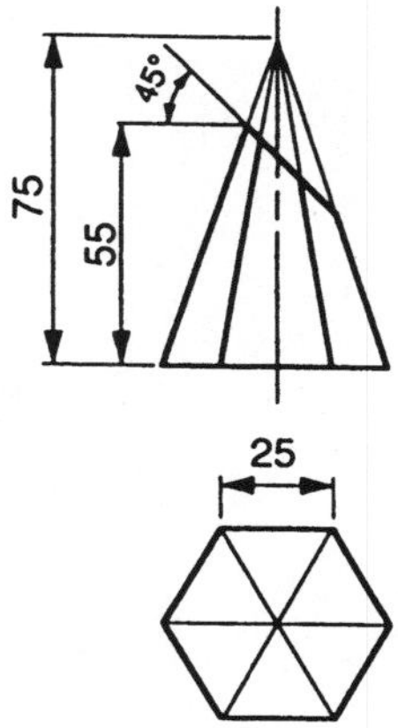

Fig 6.47

2.9 Interpenetrations

When two pipes or prisms intersect, they meet along a line called the **line of interpenetration** or **line of intersection**. The shape of this line will depend on the shapes of the intersecting pipes or prisms. For flat-sided surfaces, the line of interpenetration will be made up of **straight lines**, whereas for cylindrical surfaces the line is generally a **curve**.

Intersection of two square pipes

A front view and a top view of two square pipes intersecting at right angles, together with their line of interpenetration, are shown in fig 6.48.

Step 1: Draw the front view and top view and number the four corners of the horizontal pipe.

Step 2: The points in which the corners of the horizontal pipe meet the vertical pipe are found in the top view and projected upwards to intersect the corresponding corners in the front elevation.

Step 3: The points of intersection are joined by straight lines.

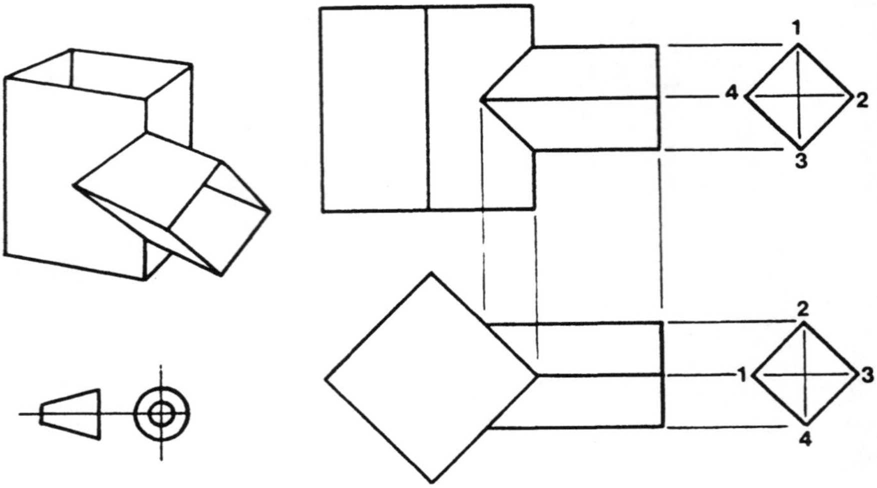

Fig 6.48

Intersection of a hexagonal pipe

Fig 6.49 shows two views of a hexagonal and a square pipe intersecting at right angles, together with their line of interpenetration.

Step 1: Draw the front view and the top view and number the six corners of the horizontal pipe, as shown.

Step 2: The points in which the corners of the horizontal pipe meet the vertical pipe are found in the top view and then projected downwards to intersect the corresponding corners in the front elevation.

Step 3: The points of intersection are joined by straight lines.

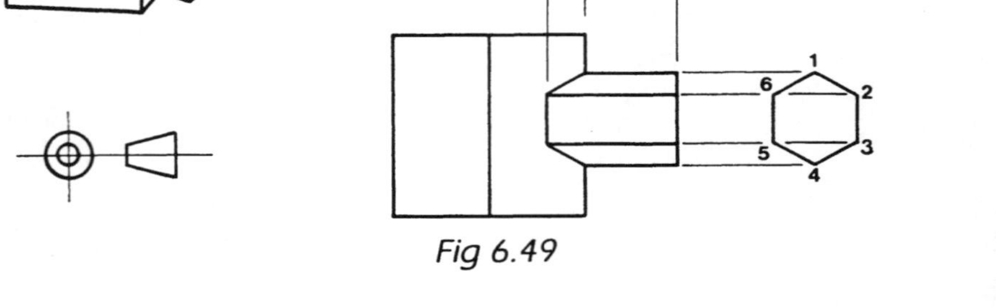

Fig 6.49

Intersection of two cylinders

Fig 6.50 shows two views of the cylinders intersecting at right-angles with their curves of interpenetration.

Step 1: Draw the front view and the top view. Divide the circumference of the top view into 12 equal parts and number as shown. Find the corresponding points in the front view. Project these lines horizontally in the top view, as shown.

Step 2: Project the points vertically downwards and where the corresponding lines meet, you will find the points of intersection.

Step 3: The points of intersection so obtained are joined by a smooth curve.

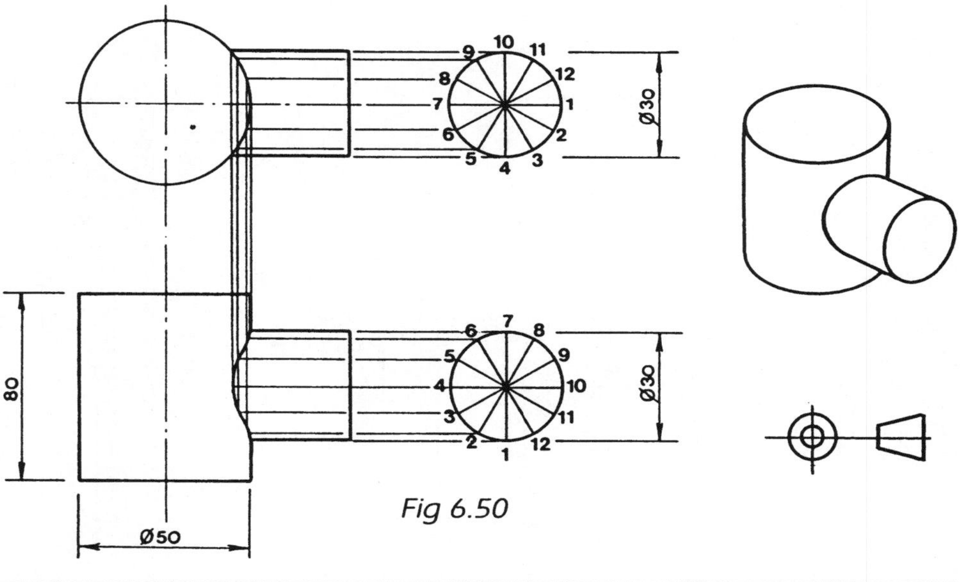

Fig 6.50

Activity 8

1. The front view and top view of two square pipes intersecting at right angles are shown in fig 6.51. Construct full scale the views, as given, and complete the front view by adding the line of intersection.
2. Two views of a horizontal square pipe intersecting a vertical cylindrical pipe are shown in fig 6.52. Copy the views and complete the front view.
3. The front view of a triangular pipe intersecting a vertical cylindrical pipe is shown in fig 6.53. Draw the front view, add a top view and complete the front view by adding a line of interpenetration.

4 Fig 6.54 shows a vertical pipe intersected by two branch pipes. Obtain the lines of intersection. Construct the views, as given, and add the line of interpenetation. Also develop the vertical pipe.

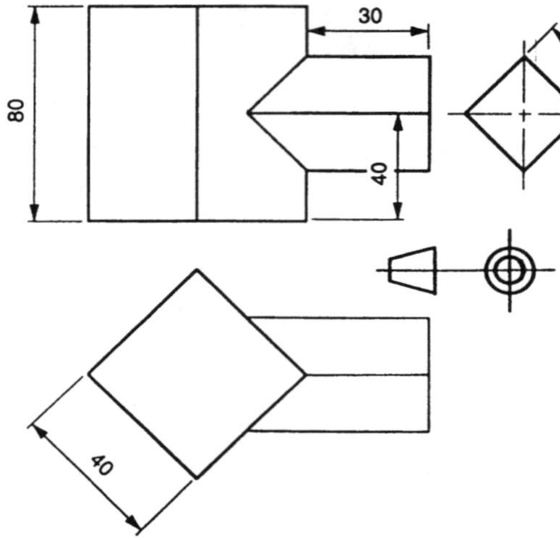

Fig 6.51

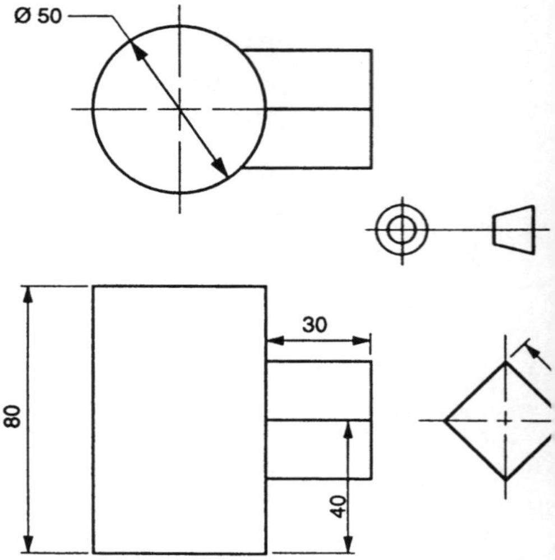

Fig 6.52

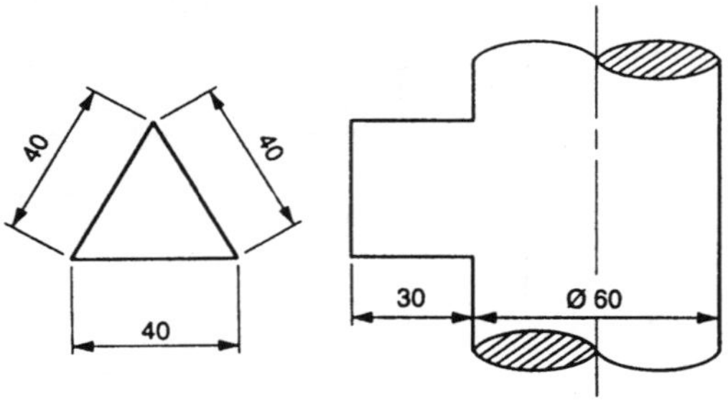

Fig 6.53

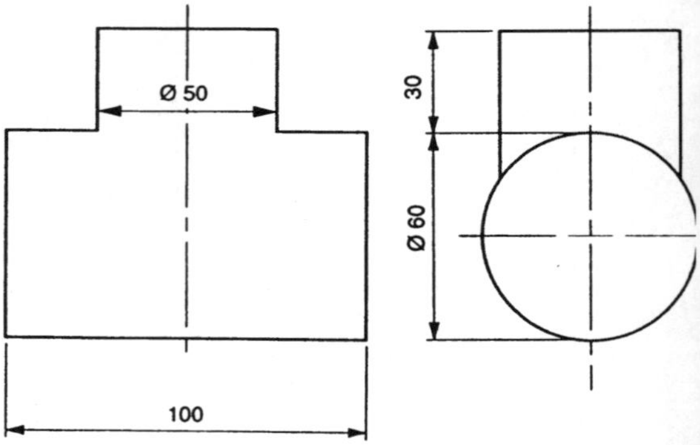

Fig 6.54

3 Summary

This unit on the layout of drawings is very important and contains crucial drawing principles that must be mastered. You must be able to identify the different categories of drawings, their basic functions and the different procedures their specific layouts require. By mastering the contents of unit 6 and referring to the previous units when required, you will be in a better position to understand and apply the knowledge and skills contained in the remaining units.

Self-evaluation

1 What is a pictorial drawing? Name three types.

2 What is meant by first-angle orthographic projection? Use sketches to support your answer.

3 What is meant by third-angle orthographic projection? Use sketches to support your answer.

4 Why are projection symbols used?

5 What is the difference between first-angle orthographic projection and third-angle ortographic projection?

6 With what types of lines will you use to construct a third view when two views are given?

7 For the layout of drawings, we use the 3-step method. Explain the 3-step method.

8 How many planes do you see on an isometric drawing?

9 Do you see any **true views** on an isometric drawing?

10 Is the isometric circle the same as an ordinary ellipse?

11 Do you see any **true views** on an oblique drawing?

12 What is the length on the 45° side of an oblique drawing?

13 Is the circle in an oblique drawing a true circle or an ellipse?

14 What is a prism?

15 What is a pyramid?

16 What is a development?

17 What is the meaning of truncated?

18 What is a truncated cone?

19 What is an example of an interpenetration line?

Answers

1 A pictorial drawing is how you see the object including the hidden detail. (a) oblique, (b) perspective, (c) isometric

2 First-angle orthographic projection means you see three views separately as shown below.

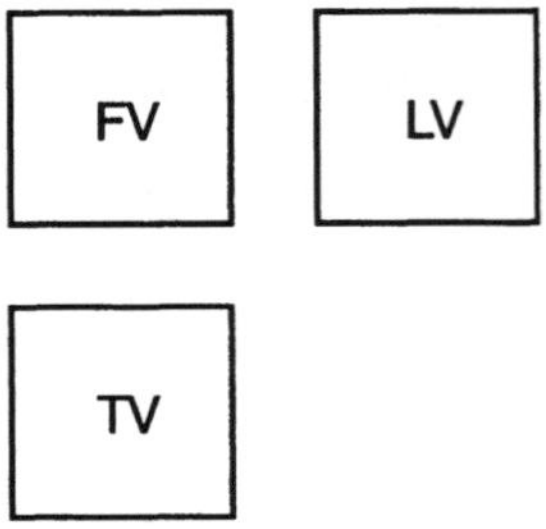

FV = FRONT VIEW

TV = TOP VIEW

LV = LEFT VIEW

3 In third-angle orthographic projection, you see three separate views, but the arrangement of the views is different as shown below.

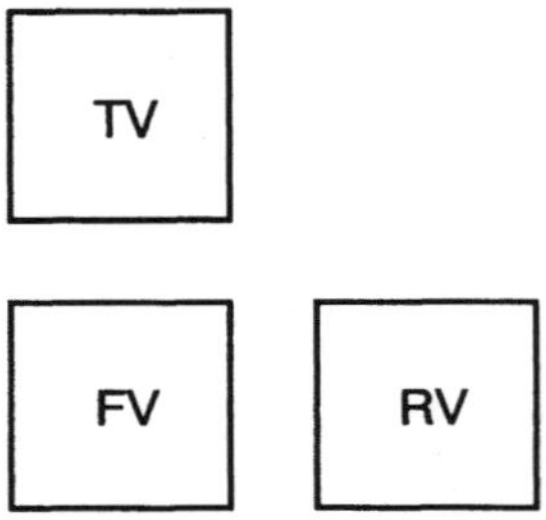

TV = TOP VIEW

FV = FRONT VIEW

RV = RIGHT VIEW

4 To see if the drawing is drawn in first- or third-angle projection.

5 The main difference is that you see first-angle orthographic projection in the **first** quadrant of the quadrant system. Third-angle orthographic projection is seen in the **third** quadrant.

6 Construction lines (projection lines) and a 45° line (mitre line).

7 This is a method you follow to set out a drawing. You start with the centre lines and build it up, step by step, until you have the full drawing.

8 Three views.

9 No, because you look from an angle onto the views.

10 No. It looks the same, but the construction is not the same.

11 Yes, in the front view.

12 It is half the usual size in order to obtain good proportions.

13 It is a true circle on the front view only.

14 A **prism** is a solid having two flat ends of the same shape.

15 A **pyramid** is a solid having a flat base of any shape and a sharp point called the **apex** or **vertex.**

16 A development means you develop a surface to form a pattern on a sheet-metal plate, which is to be cut out.

17 Truncated means an object is cut off at one end.

18 It is a cone with the top cut off.

19 When two pipes are welded together, the seam of weld is an interpenetration line.

As in the previous units, the eight activities in this unit will help you to sharpen your drawing skills. Remember to keep on referring back to all relevant information contained in the previous units. See activity 2 for further suggestions on how to check your activities yourself.

Unit 7

Sectioning

Study objectives

After studying this unit, you should be able to **draw the following**:

- ◊ Outside and sectional views of the objects
- ◊ Sectional cutting planes
- ◊ Various aspects of sectioning and hatching
- ◊ Exceptions such as ribs, shafts and bolts
- ◊ Conventional breaks in rectangular sections, round solid sections, pipes or tubing, wood and long breaks
- ◊ Conventional hatching other than for steel
- ◊ Conventional representations
- ◊ Sectional drawings
- ◊ Sectional detail drawings.

1 Introduction

This unit deals with the question of how objects which have to be drawn can be sectioned off so that they can be understood more easily.

An outside view of simple objects shows the hidden detail by means of hidden lines, called dashed lines, as explained in unit 2 (see section 2.9). This means short broken lines, which, if used too much, will cause confusion. We must therefore make use of **sections.** The function of sectioning is to show **hidden detail** by making an **imaginary cut** through an object. **These cutting surfaces** are represented by hatching. There will be **no hatching lines** when cutting through air. In other words, when you cut through a hole, for example, you will see **no** hatching lines. Certain objects, when hatched, lose their identity and should therefore not be hatched, but you must nevertheless make an imaginary cut.

For hidden lines or dashed lines refer to unit 2, fig 2.10.

For hatching lines or sectioning lines refer to unit 2, fig 2.11.

2 Sectioning

2.1 Terminology

Before we start with sectioning, you must familiarise yourself with certain terms.

Outside view

This is a view that you see from the outside. No cut is made and the hidden detail is shown by means of dashed lines. This view can represent the following views: front view, left view, right view and top view.

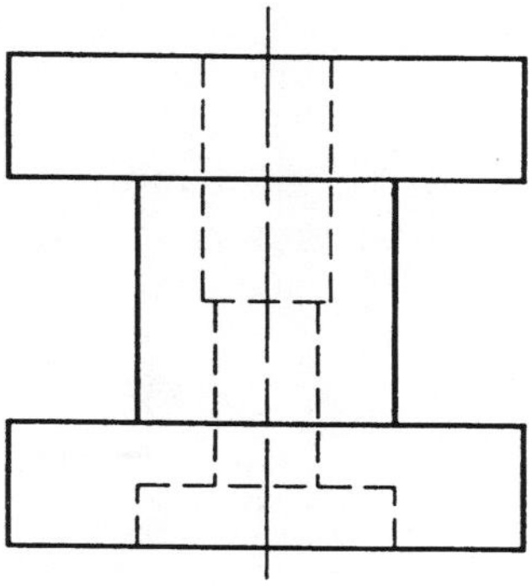

Fig 7.1

Full sectional view

This view **cuts through the centre** and must be fully hatched when you cut through material. Open spaces and holes are not hatched. This view can be a full sectional front view, a full sectional top view or a full sectional left or right view.

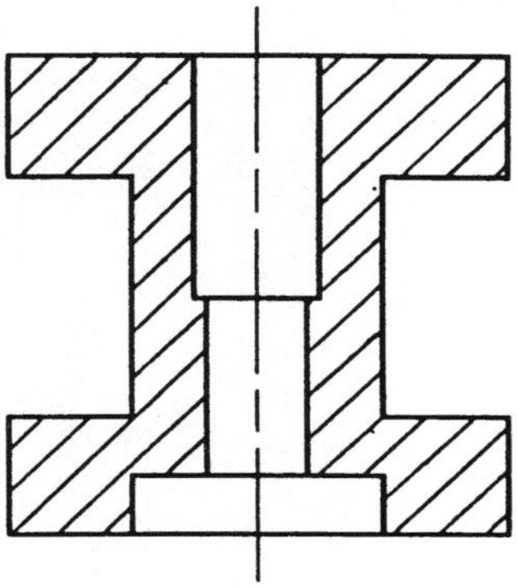

Fig 7.2

Half sectional views

Fig 7.3 is an example where only the one half is hatched and the other half is an outside view. This view is called a half sectional front view with the **left half in section**.

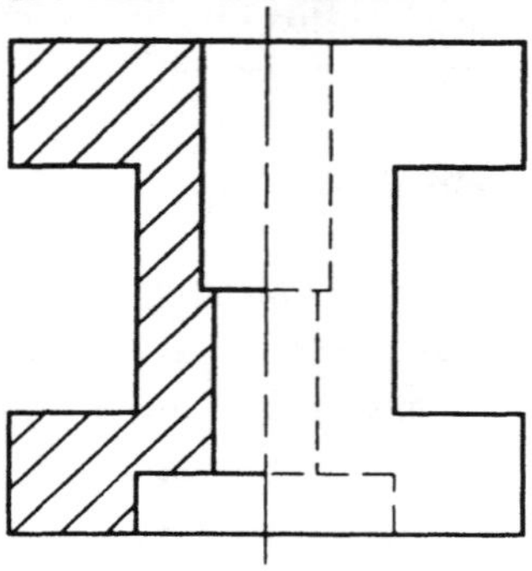

Fig 7.3

Fig 7.4 is called a half sectional front view with the **right half in section.** This is also applicable to the left view, right view and top view.

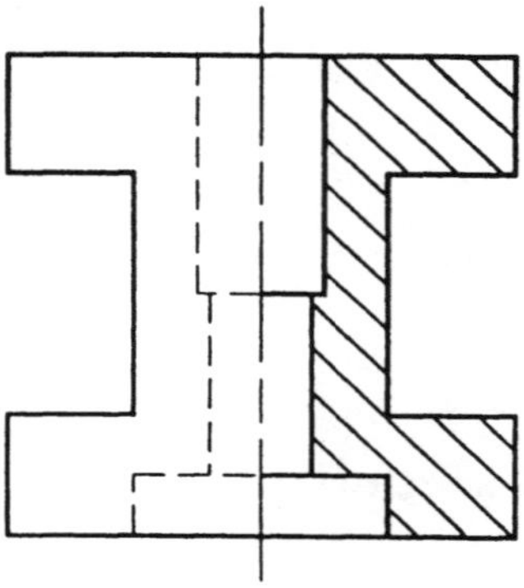

Fig 7.4

Fig 7.5 is called a half sectional front view with the **top half in section.** This will be applicable to the left view, right view and top view.

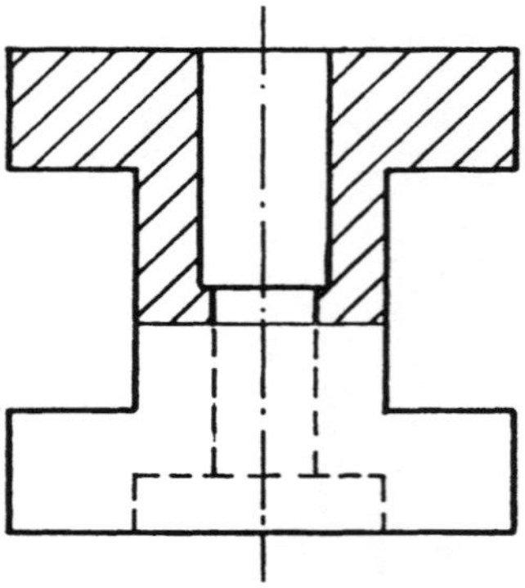

Fig 7.5

Fig 7.6 is a half sectional front view with the **bottom half in section**. This is applicable to the left view, right view and top view.

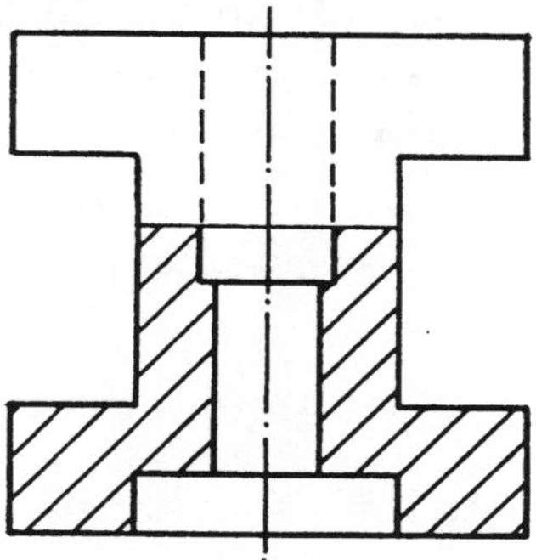

Fig 7.6

Where an area of hatching touches the centre line, the centre line must be changed to a continuous 0,3 mm thick line.

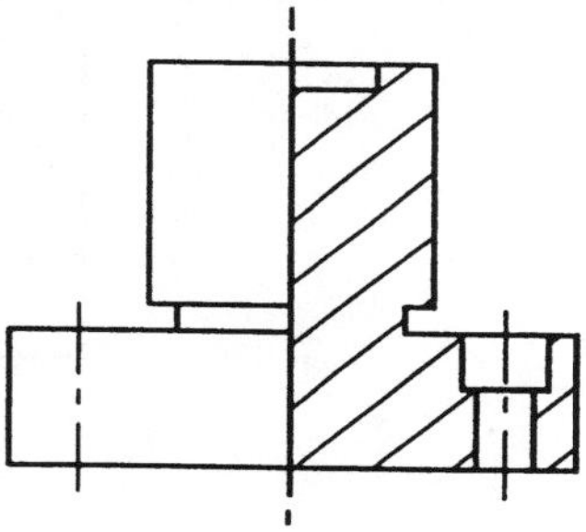

Fig 7.7

2.2 *Sectional cutting planes*

We discussed the line work of the cutting plane or sectional plane in unit 2 (see fig 2.12a). A cutting plane line is a line through an object to show where you want the object to be cut to see the hidden detail.

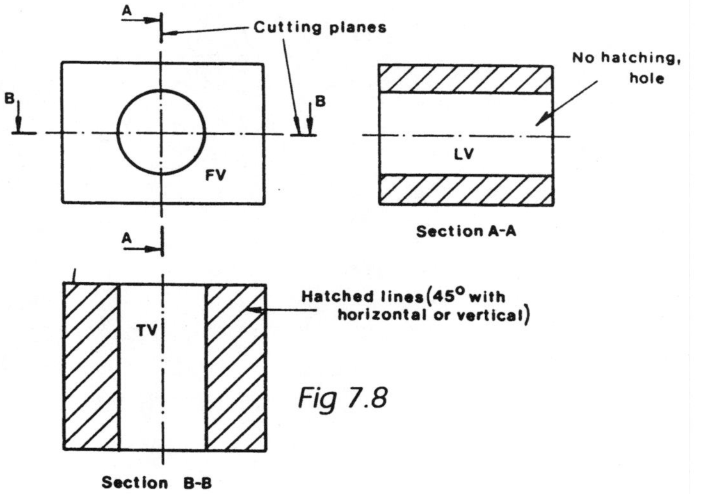

Fig 7.8

The position of the cutting plane will always be indicated. Remember, no hatching for open spaces! Notice that in each case the arrowheads on the cutting planes indicate the direction that you look from. You also probably noticed that hatching lines have to be drawn at 45° to the horizontal or vertical.

2.3 Various aspects of sectioning

Angle of sectional or hatching lines

Hatching lines should be 45° to the horizontal.

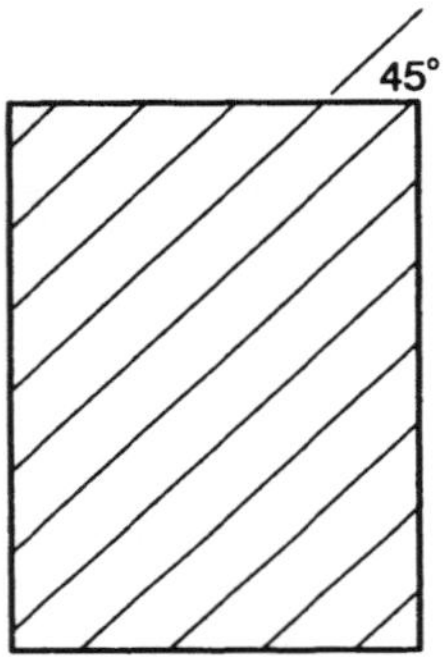

Fig 7.9

In cases where the outline of an object is at or near 45°, the hatched lines should be drawn at some other angle.

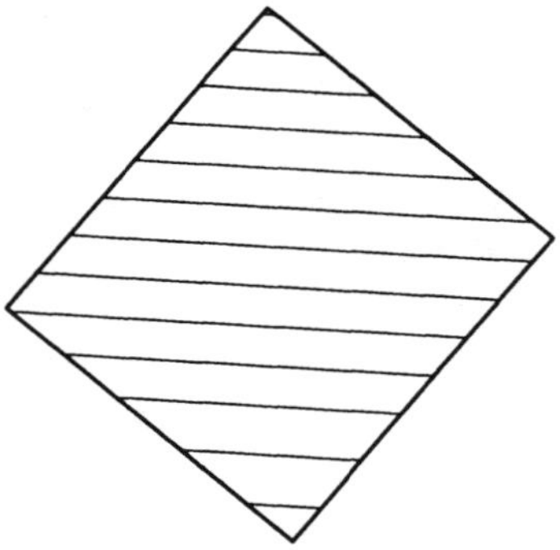

Fig 7.10

Adjacent parts

Where **two** objects are bolted together or are **adjacent** to each other, they are cut by a cutting plane. The hatching pattern in the adjacent objects must be different, as shown in fig 7.11.

Fig 7.11

You would apply the same principle in the case of **three** adjacent parts.

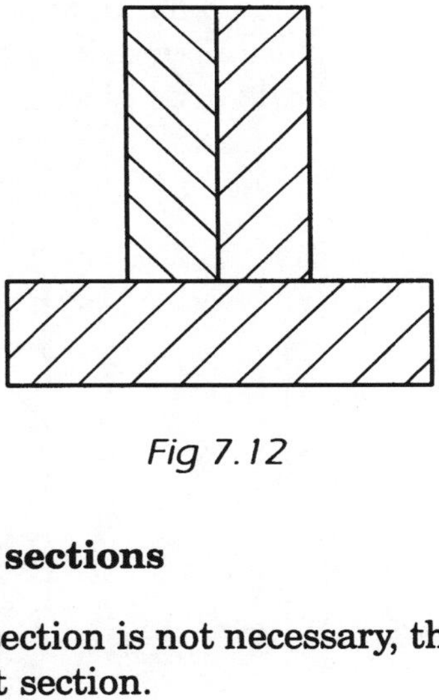

Fig 7.12

Broken or partial sections

When a full or half section is not necessary, the hidden detail may be exposed by a part section.

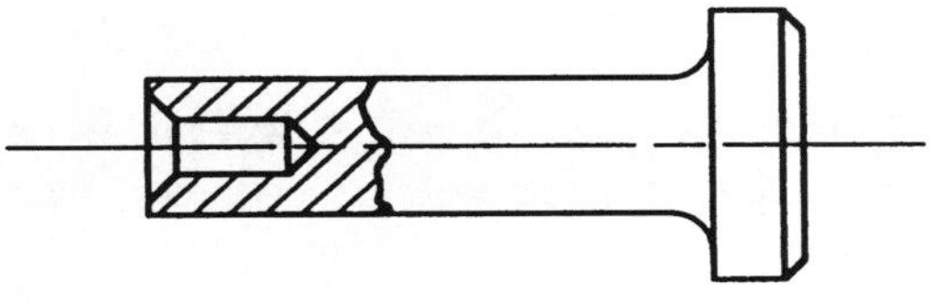

Fig 7.13

Exceptions

The parts that are not sectioned are the following:

- ◇ When the cutting plane passes **longitudinally** through a solid part that has no hidden detail, the part is not hatched.
- ◇ Bolts, nuts, screws and rivets.
- ◇ Keys, cotters, pins, shafts and spokes. Cotters are bolts or wedges for securing parts.
- ◇ Another common exception occurs when the cutting plane cuts a **rib longitudinally** as shown in fig 7.14a. The false impression of thickness suggested by the sectioning in fig 7.14b is avoided by leaving the rib unsectioned, as shown in fig 17.14c, which gives a truer picture of the web. In this case the **web** is the triangular-shaped piece of material which connects the vertical with the horizontal pieces.

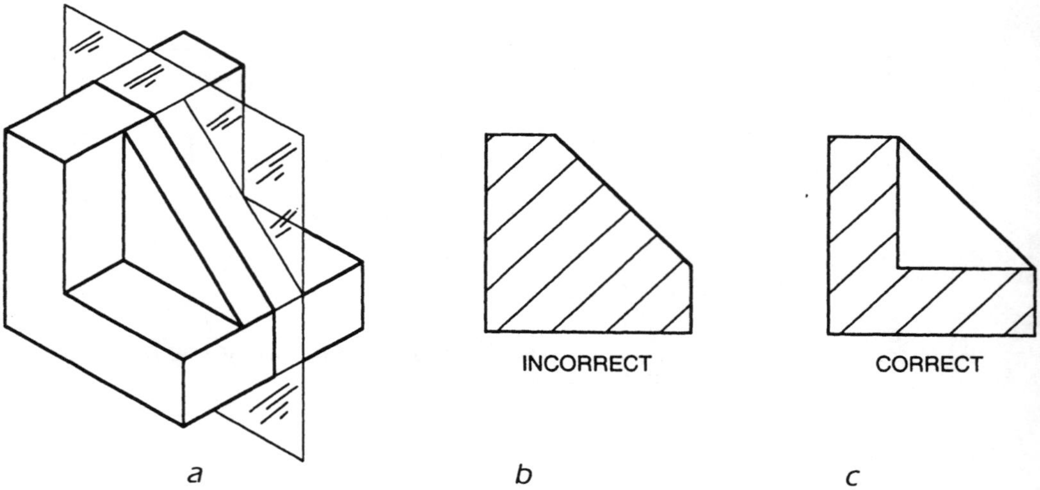

Fig 7.14a–c

On the other hand, if the rib is cut **transversely** as shown in fig 7.15a, the rib will be sectioned as shown in fig 7.15b.

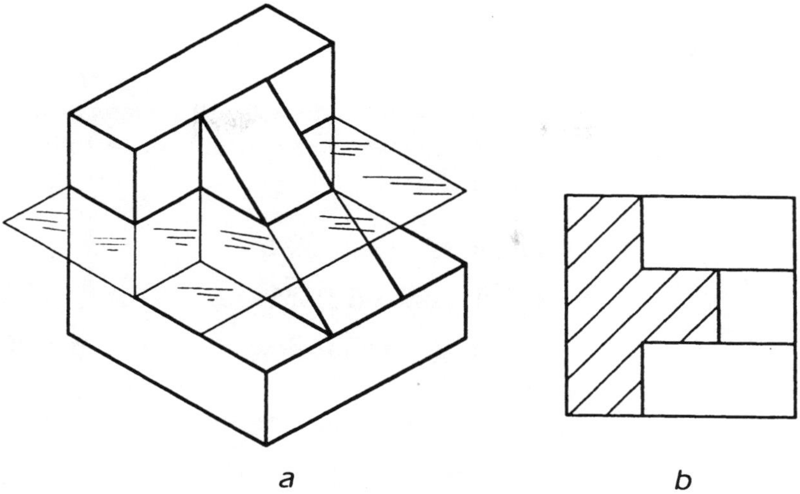

Fig 7.15a–b

A solid shaft and a bolt are not hatched longitudinally but are hatched crosswise.

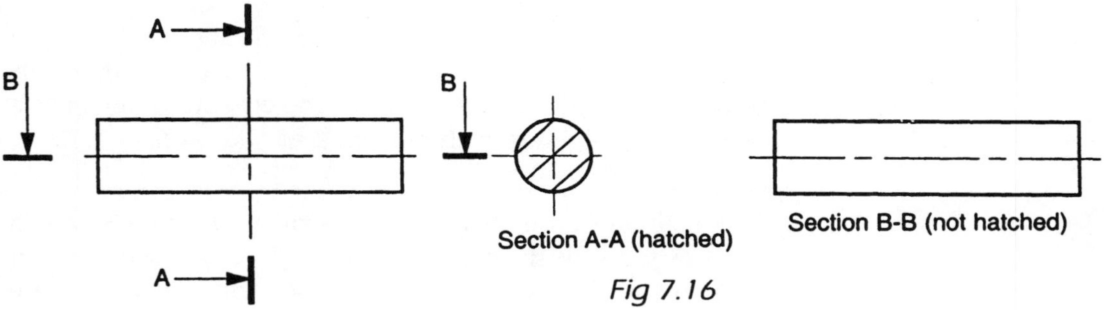

Fig 7.16

Conventional hatching of materials other than steel

Table 7.1 shows how material other than steel is represented when sectioned.

Table 7.1

MATERIAL	HATCHING
Thermal insulation	
Glass	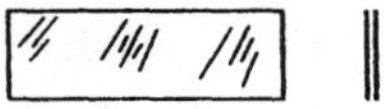
Wood	

MATERIAL	HATCHING
Concrete	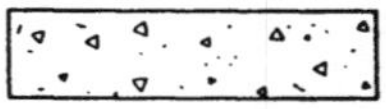
Liquids	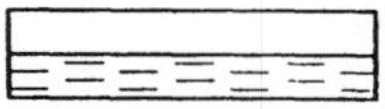
Brickwork	

2.4 How to draw a sectional view

Two or three views of an object will always be given and you will be assessed on whether you can draw the following according to the code of drawing:

- ◇ An outside front view
- ◇ A full sectional front view
- ◇ A sectional front view with the top half or bottom half in section
- ◇ An outside left or right view
- ◇ A full sectional left or right view
- ◇ A sectional left or right view with the bottom half or top half in section
- ◇ An outside top view
- ◇ A full sectional top view.

Only the cutting planes will be shown and you must draw the section views accordingly. The arrows on the cutting planes show the directions from which one should look at the section.

Activity 1

1 Fig 7.17 shows an outside front view and an outside top view of a bearing. Draw in first angle orthographic projection the following three views of the bearing. Scale 1 : 1.

  a The outside front view

  b A sectional left view on cutting plane C-C

  c A sectional top view on cutting plane D-D.

Dimension the views and insert the title, scale and projection symbols. Two views are given and you must find the third view. Refer to unit 6 if you are not sure how to go about this.

Once you have drawn the sections, dimensioned the views and inserted the title and scale, you can check your answer by referring to the answers section of this unit. The solution (fig 7.23) is at the end of the unit to motivate you to try and construct the correct drawing on your own before you look at the answer.

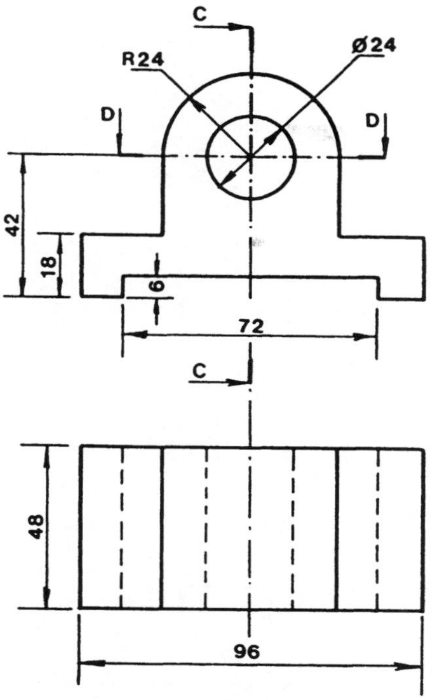

Fig 7.17

2 Fig 7.18 shows two views in first angle orthographic projection of a V-block. Draw, in first angle orthographic projection and full scale, the following three views of the V-block:

- **a** A sectional front view on A-A
- **b** The outside top view
- **c** A sectional left view on cutting plane B-B.

Insert the dimensions and the title.

3 Fig 7.19 shows two views in first angle orthographic projection of a block. Draw in first angle orthographic projection and full scale the following three views of the block:

- **a** A sectional front view on cutting plane A-A
- **b** A sectional left view on cutting plane B-B
- **c** The outside top view.

Insert the dimensions and the title.

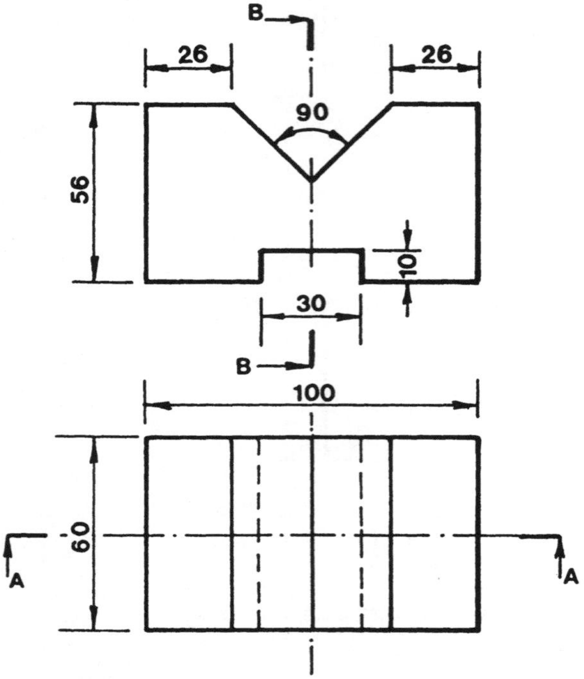

Fig 7.18

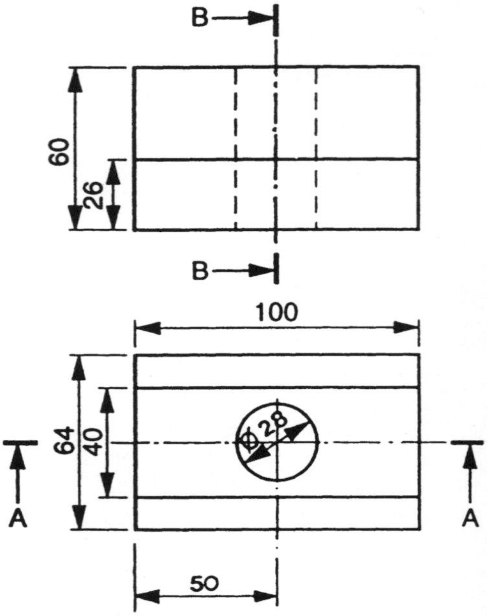

Fig 7.19

2.5 Sectional detail drawings

Detail drawings are working drawings of real objects. They show all the details necessary to manufacture the object. So far in this unit we have discussed everything concerning sectioning. Detail drawings must also be sectioned for clarity where applicable. In the next unit we will be dealing with assembly drawings, which show how the various parts of an assembly fit together. You will be required to be able to do detail drawings in order, for example, to draw a sectional assembly drawing.

A detail drawing must consist of the following:

- ◇ Dimensioning, correct linework, correct scale, etc.
- ◇ Threads and nuts or bolts where applicable
- ◇ Machining symbols and tolerances where applicable.

Fig 7.20 is a typical detail drawing with all its details and specifications.

CRANK SHAFT
SCALE = 1 : 2

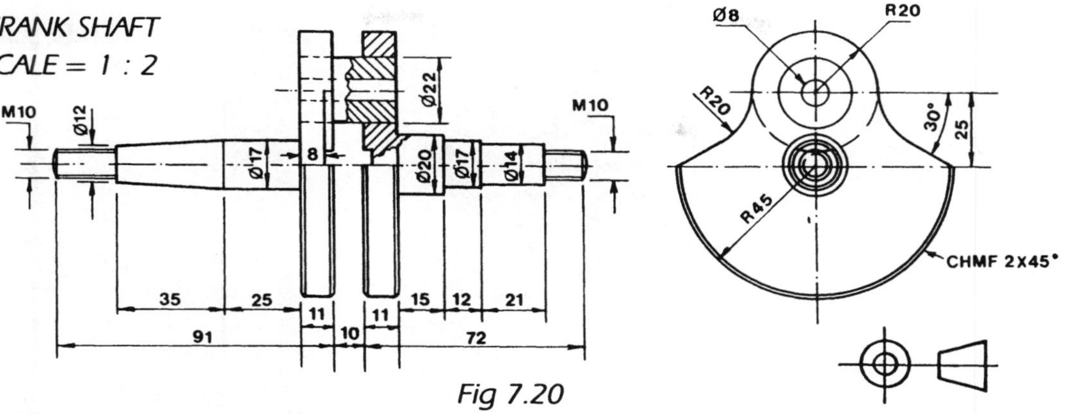

Fig 7.20

Activity 2

1 **Fig 7.21 shows a half sectional front view on cutting plane A-A and an incomplete outside side view of a V-pulley. Draw, full size, in first angle orthographic projection, the following two views of the pulley:**

a **A full sectional front view on cutting plane A-A**

b **A complete outside left view.**

Dimension the two views and print the title and scale.

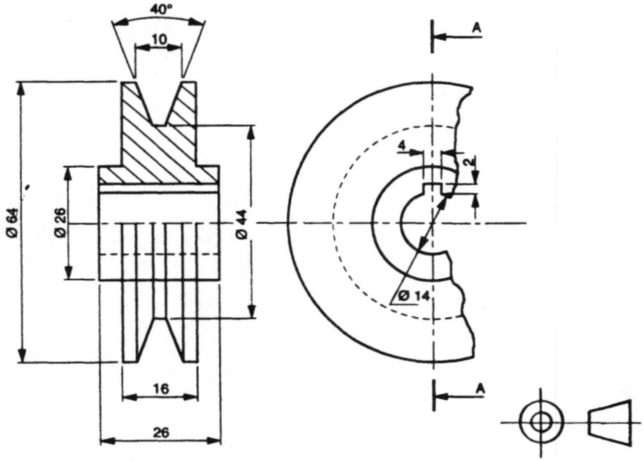

Fig 7.21

2 **Fig 7.22 shows two outside views of a clamp. Draw, to a scale of 1:5, in first angle orthographic projection, the following view of the clamp:**

a **A half-sectional front view on cutting plane B-B with the right half in section**

b A sectional left view on cutting plane A-A

c An outside top view.

Dimension the two views and print the title and scale:

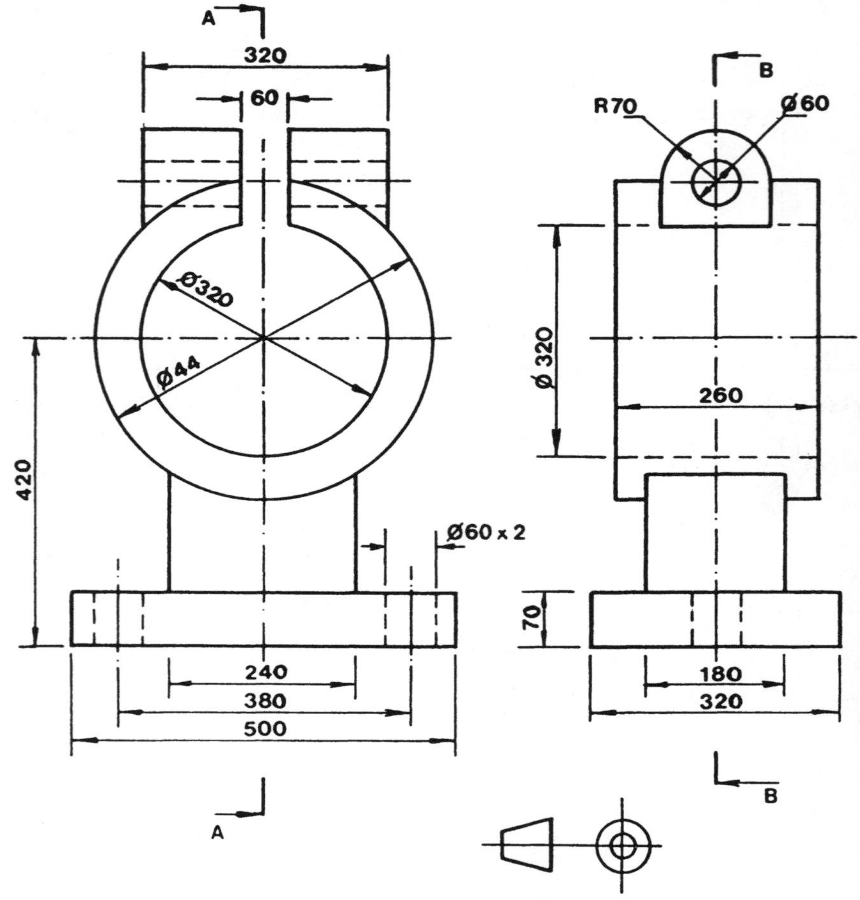

Fig 7.22

---

3 Summary

Sections are very important in engineering drawing because when you have a number of hidden details, you will make an imaginary cut through the object to see or describe what is going on inside that specific object. In this unit you have been supplied with the necessary information and common practices so that you can draw the different types of sections, interpret the cutting planes and explain the reasons for the arrows on the cutting planes. We have also discussed which objects we are going to show in sections.

Self-evaluation

1 Do you see any hatched lines when cutting through a hole?
2 What is an outside view?
3 How many half sectional views are there? Name them.
4 What is a cutting plane line?
5 Are the hatched lines in the same direction for adjacent parts?
6 What is a partial section?
7 Do you show a solid shaft and bolt cut longitudinally in section?
8 What is a detail drawing?
9 Do you show a rib in section when it is cut longitudinally? Give a reason for your answer.
10 Do you show a rib in section when it is cut transversely? Give a reason for your answer.

Remember the importance of **review sheets.** At the end of the course you should have created personal review sheets for every unit. So instead of having to study 250 pages, you might need to study only 50 pages, which will make it much easier for you to prepare for the exams.

Answers

1 No.
2 It is a view that you see from the outside. No hatched lines.
3 Four.
  a Half sectional view with the left half in section
  b Half sectional view with the right half in section
  c Half sectional view with the top half in section
  d Half sectional view with the bottom half in section
4 A cutting plane line is a line through an object to show where you want the object to be cut.
5 No. You must change the direction.
6 When a full or half section is not necessary, the hidden detail may be exposed by a part section.

7 No.

8 A detail drawing shows the details and specifications required to produce a part.

9 No. A false impression of the material cut away will be created.

10 Yes. A real impression of the section will be given.

Solution to example 1

Fig 7.23 shows the completed views of the bearing shown in fig 7.17. All the answers to the questions asked in the example have been given. Make sure you are able to apply this specific method because this is the only correct method. Once you have compared your drawing with the solution, try to complete activity 1.

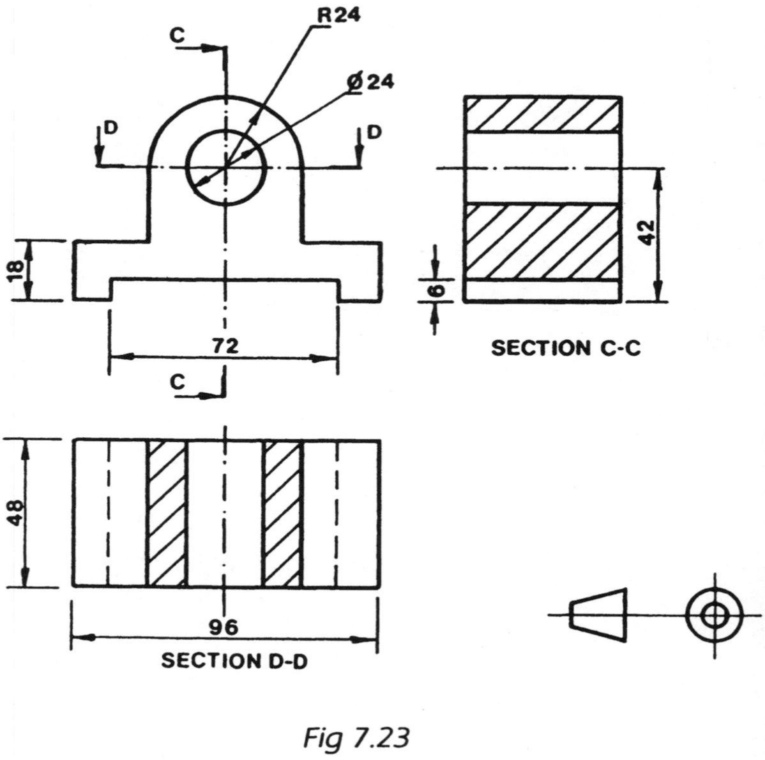

Fig 7.23

Unit 8

Conventional representations

Study objectives

After studying this unit, you should be able to interpret and draw the conventional representations of the following common features on drawings.

- ◇ Drilled and tapped holes
- ◇ Studs and hexagonal bolts
- ◇ Springs
- ◇ Breaks of large components
- ◇ Welded joints

1 Introduction

In this unit we will learn the method for showing complex features on a drawing. We simplify these features by using conventional or standard representations. Because these representations do not show the true appearance of the feature, it is very important that you draw the conventional representations accurately so that anyone reading your drawing can understand.

2 Holes and fasteners

2.1 Representation of a drilled hole

Fig 8.1a shows a typical representation of a drilled hole in a machine part.

Fig 8.1b shows the angles to use for the construction of a drawing.

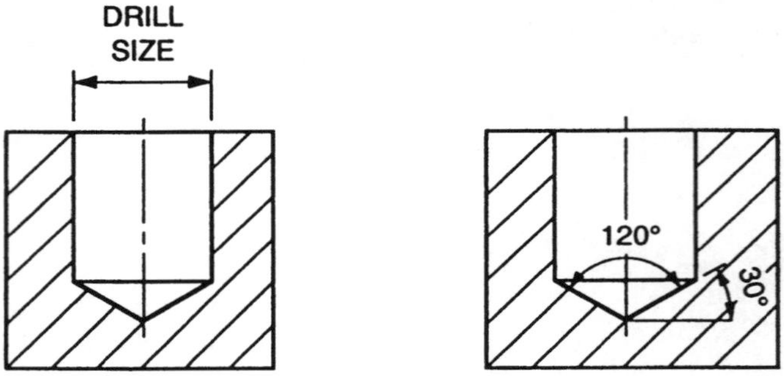

a *b*

Fig 8.1a–b

2.2 Representation of a tapped hole (threaded hole)

When you show a metric thread, you have to show the drilled hole first. The metric thread will be just a little bigger than the drilled hole.

Fig 8.2 shows a typical representation of a tapped hole in a machine part (front and top views). Metric threads can be shown as M36 or M36 × 4 where:

M indicates that it is a metric thread 36 indicates the nominal diameter in millimetres and the 4 indicates the pitch of the thread in millimetres.

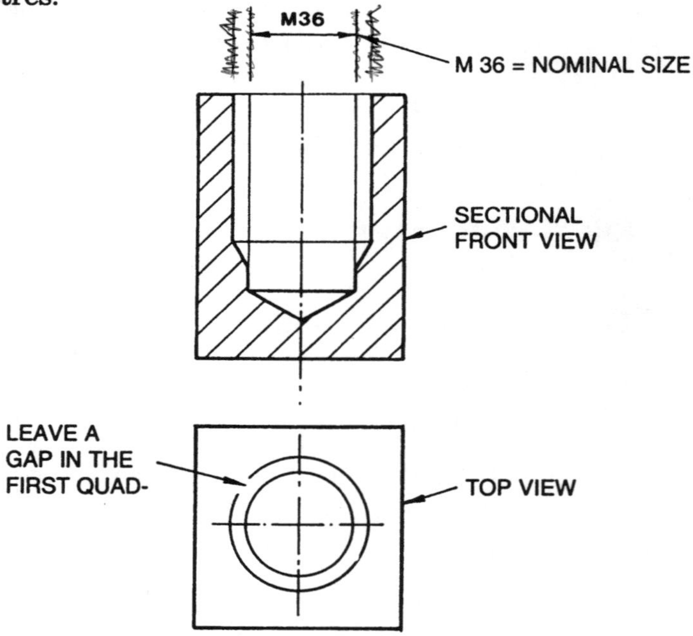

Fig 8.2

Fig 8.3 shows the different angles and measurements needed to draw a representation of a metric thread and the drilled hole. The nominal size of the metric thread is the bigger diameter and the diameter is 0.8 × nominal size (M36) in millimetres.

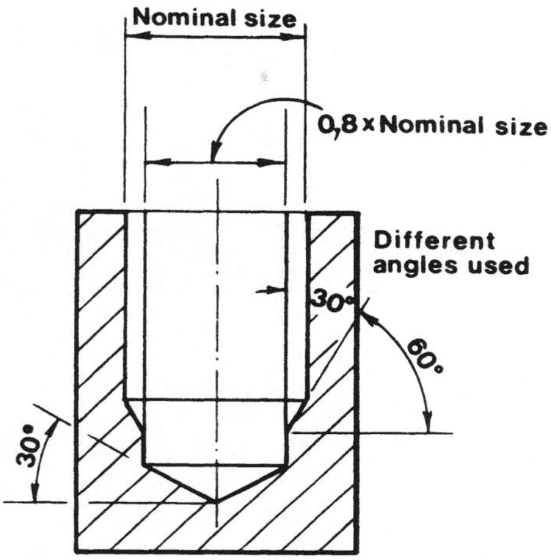

Fig 8.3

2.3 Construction of a hexagonal nut

Follow the six steps below to draw a hexagonal nut or bolt head, as illustrated in fig 8.4.

Step 1: Begin the view which contains the circles by drawing a circle to represent the diameter d of the bolt. Draw a pair of parallel lines a distance d apart above the circle.

Step 2: Draw the chamfer circle with radius $r = 0{,}8 \times d$, from step 1.

Step 3: On the outside of the chamfer circle, draw a hexagon and then project the corners onto the other two views.

Step 4: Use a 60° set-square to obtain the centres for the three arcs on the three-faced view, as well as the centres for the two arcs on the two-faced view.

Step 5: Draw the five arcs and, using a 30° set-square, chamfer the top corners of the three-faced views. Note that the corners of the two-faced view are NOT chamfered.

Step 6: Complete the views by darkening the outlines, adding centre lines and erasing all unnecessary lines.

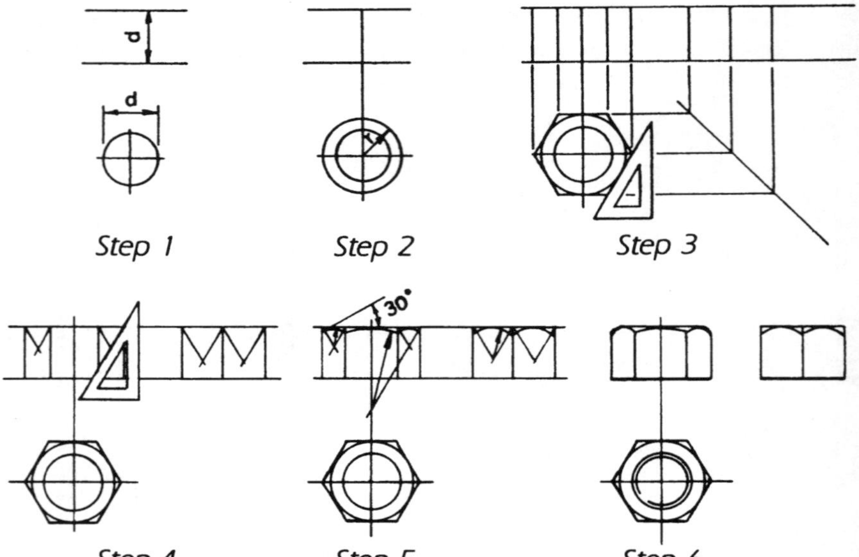

Fig 8.4

2.4 Representation of a hexagonal head bolt

Fig 8.5 shows a typical drawing of a hexagonal head bolt. The thickness of the bolt head = 0,7 × nominal size. The construction of a hexagonal head bolt is the same as for construction of a hexagonal nut. The root diameter of the thread is 0,8 × nominal size in millimeters.

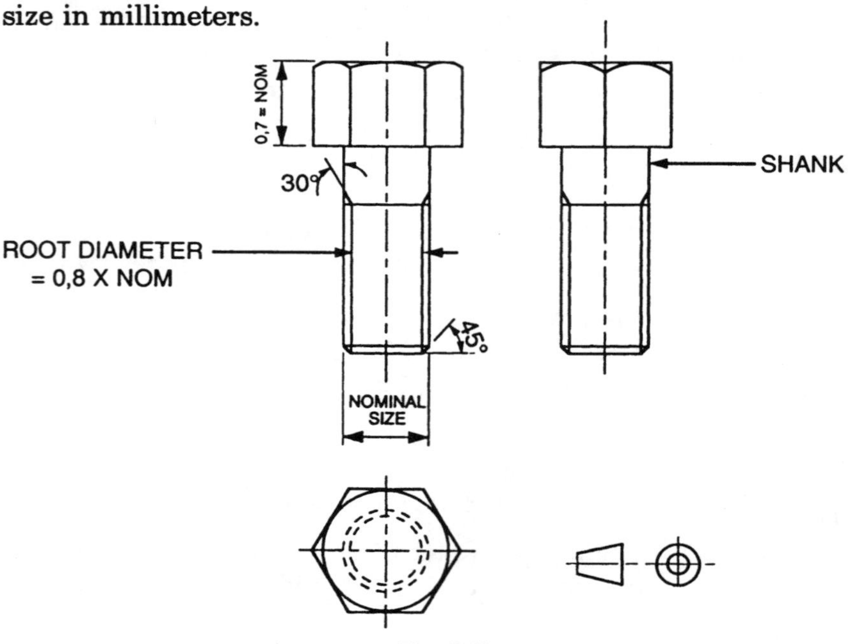

Fig 8.5

The thread of a hexagonal head screw starts just below the head. The thread of a hexagonal head bolt leaves a portion that is not threaded. This is called a shank (see fig 8.5).

2.5 Representation of a stud

Fig 8.6 shows two views of a stud. The middle portion of the stud is not threaded.

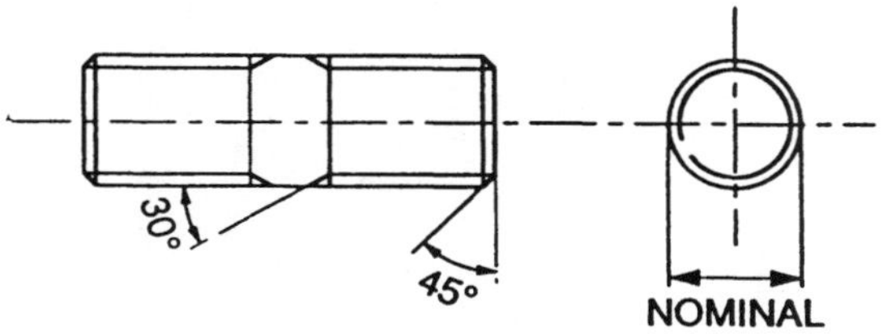

Fig 8.6

2.6 Representation of a stud assembly

The assembly shown in fig 8.7 consists of two componen ts bolted together with a stud and a nut.

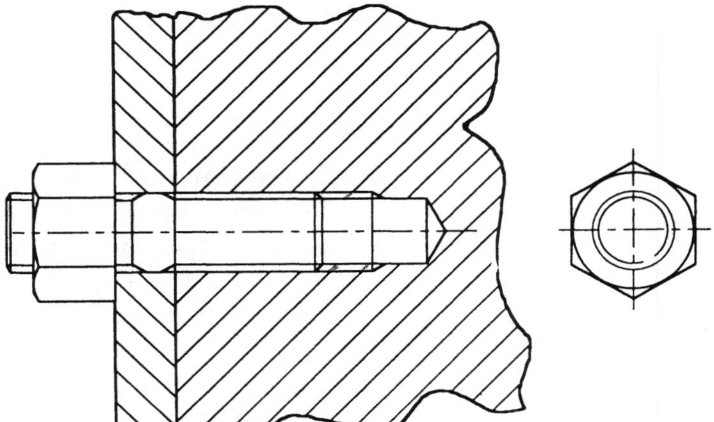

Fig 8.7

Activity 1

1 Draw, full size and according to first-angle orthographic projection, a front view, left view and a top view of the following:

   a M30 hexagonal bolt with a total length of 100 mm. Threaded length is 40. Do not insert dimensions.

   b M24 hexagonal nut

2 Draw, full scale, the given view, as shown in fig 8.8 below. Assume that the dimensions not given are in good proportion, to SABS 0111.

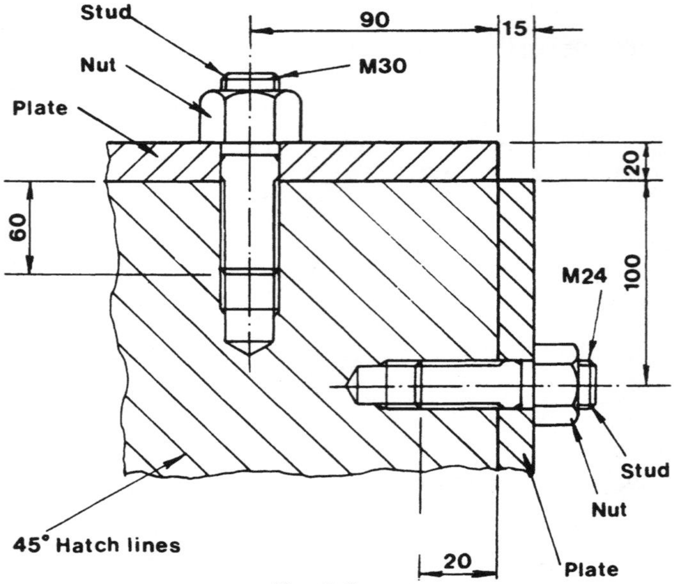

Fig 8.8

3 Springs

3.1 Representation of springs

Table 8.1 shows how different types of springs may be represented in outline and section views. The simplified representation may be used to save time when the loss of detail is not critical.

Table 8.1

Type of spring	*Outline view*	*Sectional view*	*Simplified representation*
Cylindrical helical compression spring of wire of circular cross-section	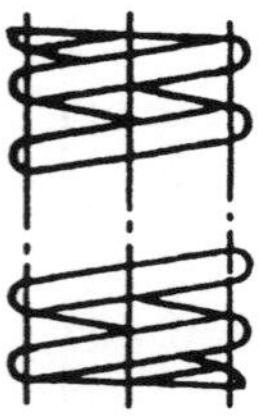	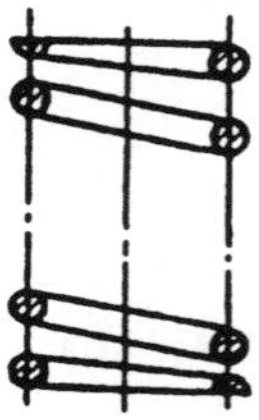	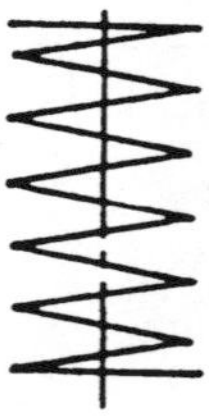

Type of spring	*Outline view*	*Sectional view*	*Simplified representation*
Cylindrical helical compression spring of wire of rectangular cross-section	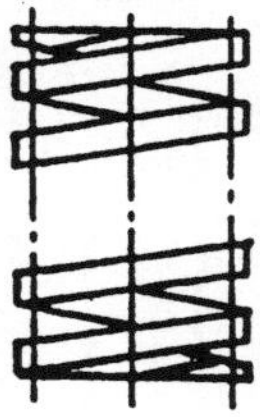	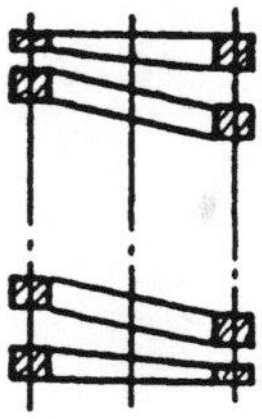	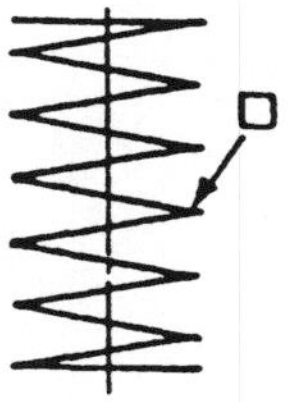
Conical helical compression spring of wire of circular cross-section		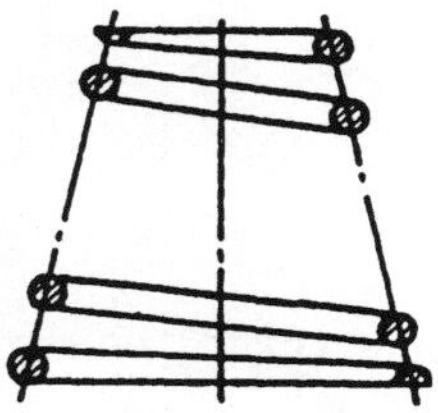	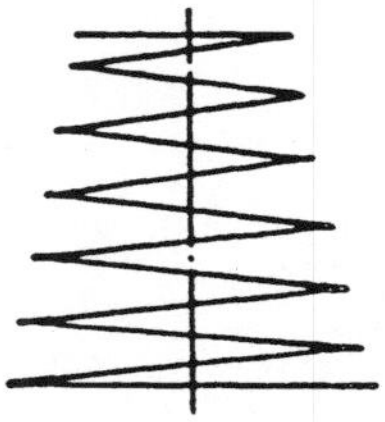
Cylindrical helical tension spring of wire of circular cross-section	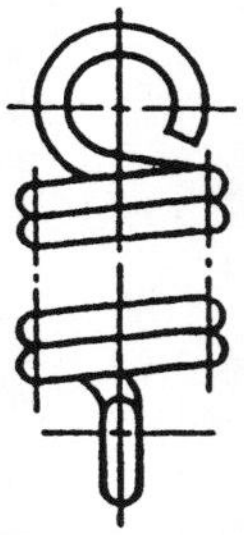	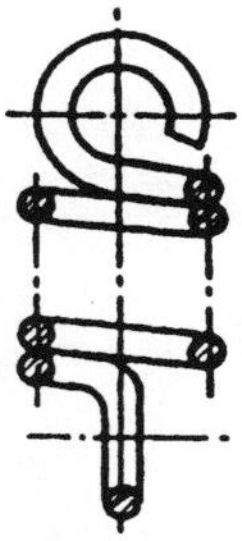	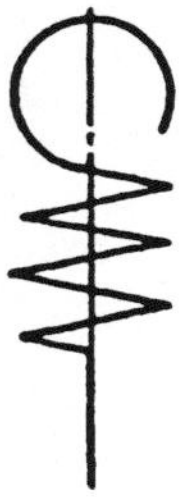
Barrel-shaped helical tension spring of wire of circular cross-section		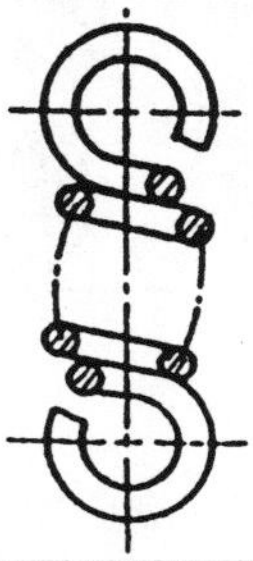	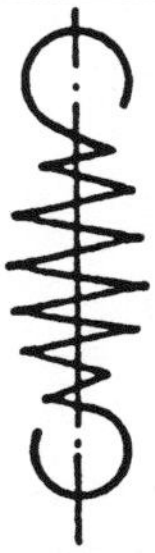

4 Breaks

A **long part** with a uniform cross-section may be drawn to fit on a standard size drawing sheet by cutting out a portion of the length.

◇ Rectangular section

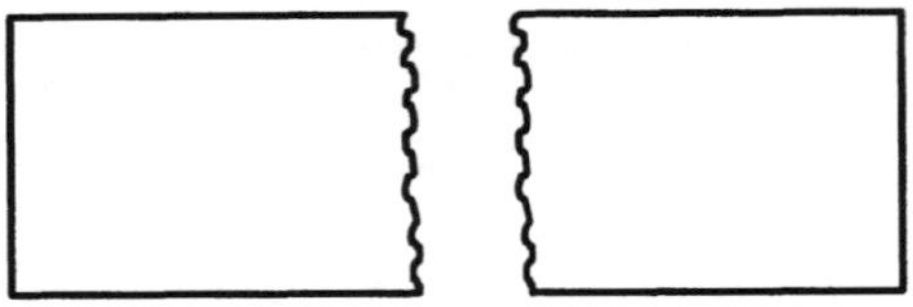

Fig 8.9

◇ Round solid section

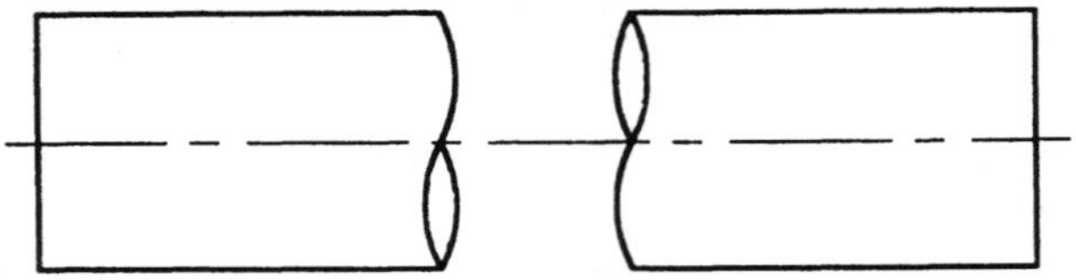

Fig 8.10

◇ Pipe or tubing

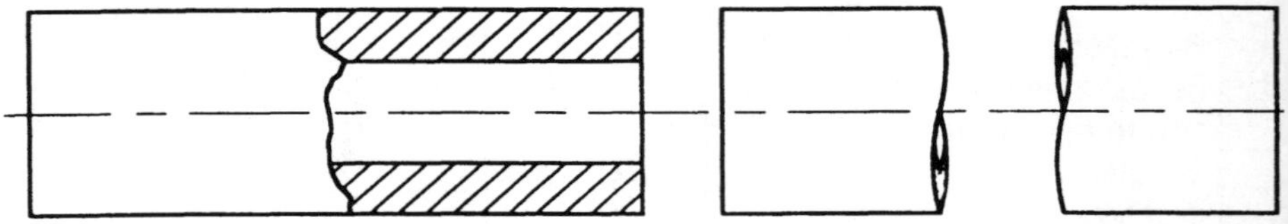

Fig 8.11

◇ Wood

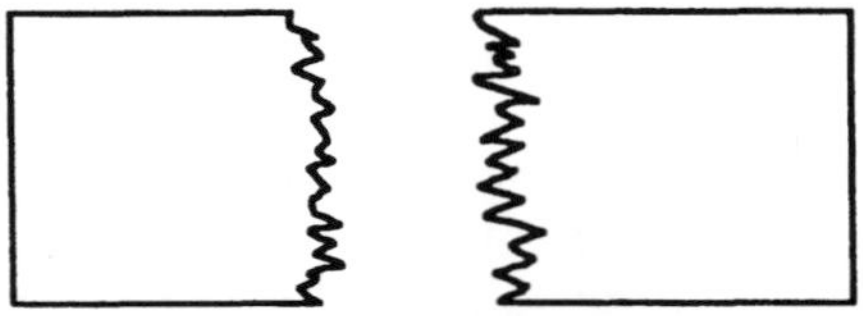

Fig 8.12

◇ Long breaks (all materials)

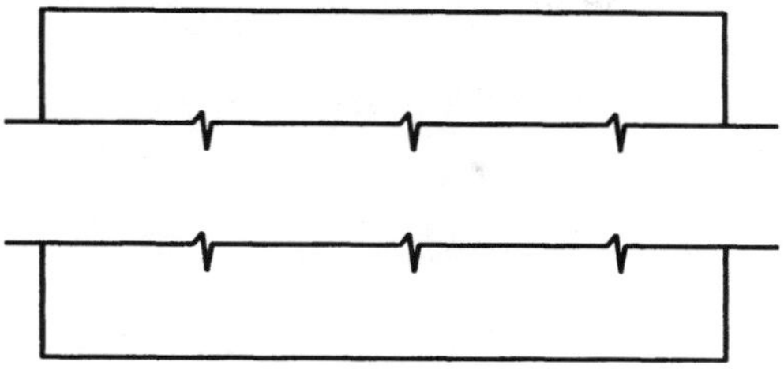

Fig 8.13

Conventional hatching other than for steel

These tables were provided for your own interest as they are mostly used in civil, electrical and chemical engineering.

5 Welded joints

Nowadays welding is used far more extensively than screws, bolts or rivets for fastening parts together. For this reason you must have a knowledge of the standard way of representing welded joints in egineering drawings. On drawings, welds themselves are not drawn but are represented by **symbols**.

5.1 Types of welded joints

There are four types of welded joints: butt joints, lap joints, T-joints and corner joints.

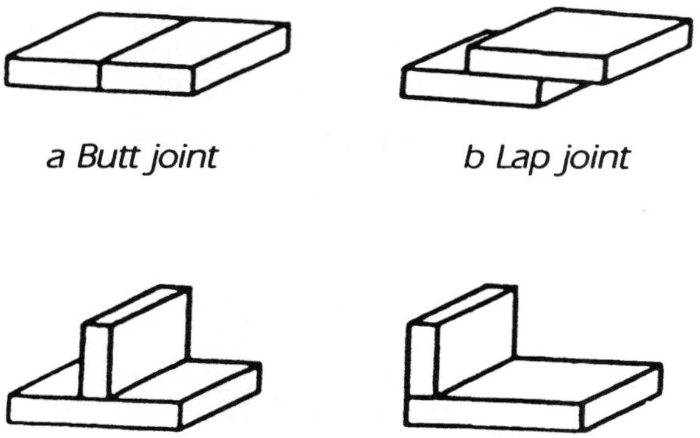

a Butt joint *b Lap joint*

c T-joint *d Corner joint*

Fig 8.14a–d Types of joints

5.2 Representation of welds

The symbolic representation of a weld has three basic elements: a reference line, an arrow and a weld symbol as shown in fig 8.15.

1 Reference line. These lines are usually horizontal.

2 Arrow.

3 Weld symbol.

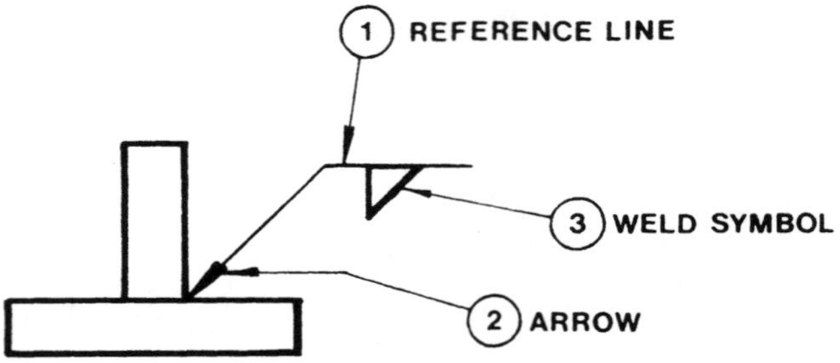

Fig 8.15

There are different ways of inserting the weld symbols on the reference line. They can be inserted below, on the top or on both sides of the reference line, indicating the side the weld is on. See the following examples:

Fillet weld on the right-hand side

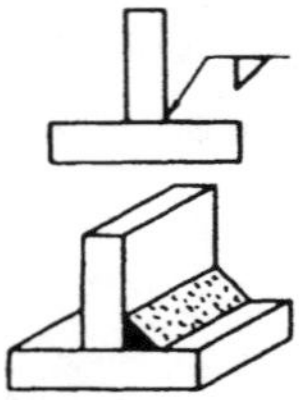

Fig 8.16

Fillet weld on the left-hand side

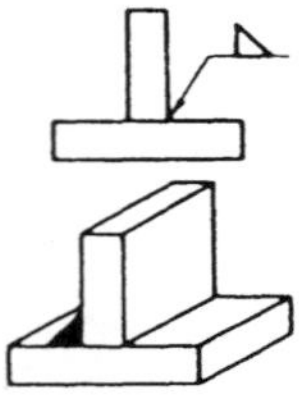

Fig 8.17

Fillet weld on both sides

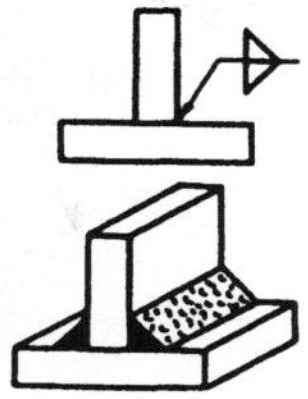

Fig 8.18

Note: weld symbol can go on either side

Single-V butt weld

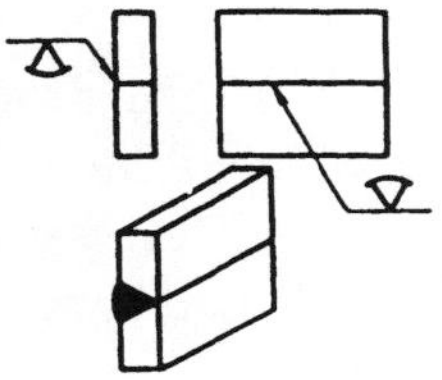

Fig 8.19

Fillet weld on a single lap joint

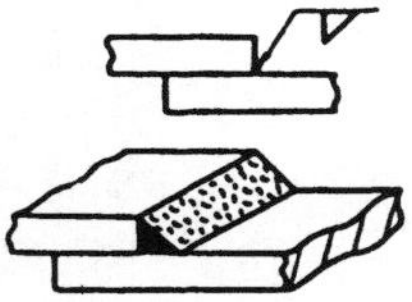

Fig 8.20

Single-V weld on a corner joint

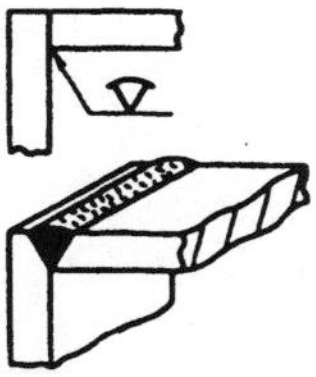

Fig 8.21

5.3 Supplementary symbols

Weld-all-around

This is indicated by a circle with its centre at the intersection of the reference line and arrow.

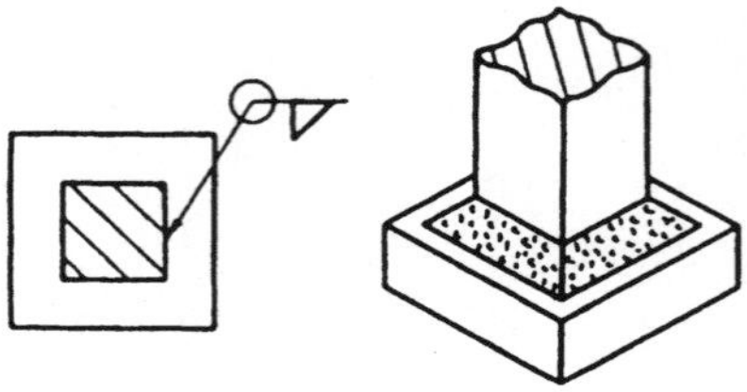

Fig 8.22

Welds to be made on site

These are indicated by a smaller filled-in circle, also with its centre at the intersection of the reference line and arrow.

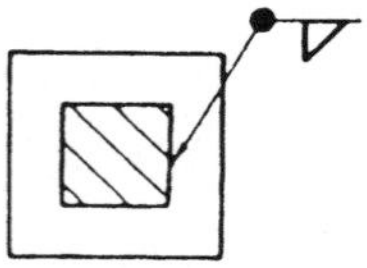

Fig 8.23

Weld-all-around on site (construction)

This symbol depicts a filled-in circle within a circle on the intersection of the reference line and arrow.

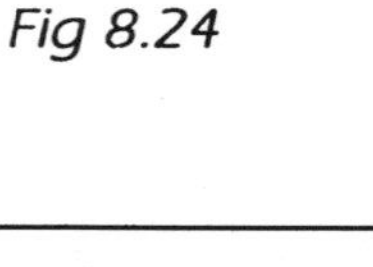

Fig 8.24

Activity 2

Prepare review sheets for the material presented in this unit. Keep it with the others and refer to it often.

6 Summary

In this unit you have learned how to draw the conventional representations of many common features. Other features, such as gear teeth and surface finish have their own conventions. You will learn about these as needed when you encounter them.

It is important that you use only the conventional representations as contained in SABS 0111, otherwise your drawing may be misinterpreted. Also, you should not create your own representations as others may not understand them.

You will need to use these representations when making assembly drawings in the next unit.

Self-evaluation

1 When you show a metric thread, do you have to show the drilled hole first? Give a reason.

2 Where do you get a tapped hole?

3 What is the meaning of the symbol M?

4 What is meant by nominal size?

5 What is the thickness of the root diameter of a thread? Just give the formula.

6 What is the thickness of a bolt head? Just give the formula.

7 Define a stud.

8 Make a sketch of a conventional break.

9 How do you show a break in a shaft?

10 How do you show a break in a hollow shaft?

11 Define welding.

12 How many types of welding joints are there?

13 What is a weld symbol?

14 What is meant by the term ‘weld all around’?

15 How do you show the symbol for ‘weld all around’?

16 What is meant by the term ‘weld on site’?

17 Show the symbol for ‘weld on site’?

18 What is a fillet weld? Show a sketch.

19 Show the symbol for a fillet weld.

20 Show the symbol for ‘weld all around on site’

Answers

1 Yes, because you must first drill the hole before you tap a thread.

2 On any machine part where assembly is necessary.

3 M means metric thread.

4 Nominal size is the overall diameter of a thread.

5 0,8 × M (nominal size)

6 0,7 × M (nominal size)

7 A rod threaded on both ends with an unthreaded portion in the middle

8

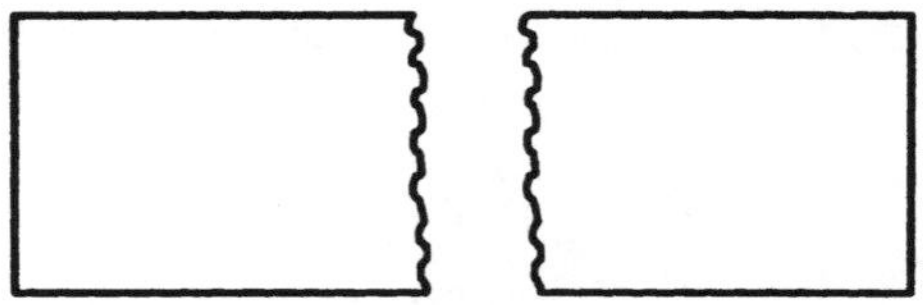

Fig 8.9

9

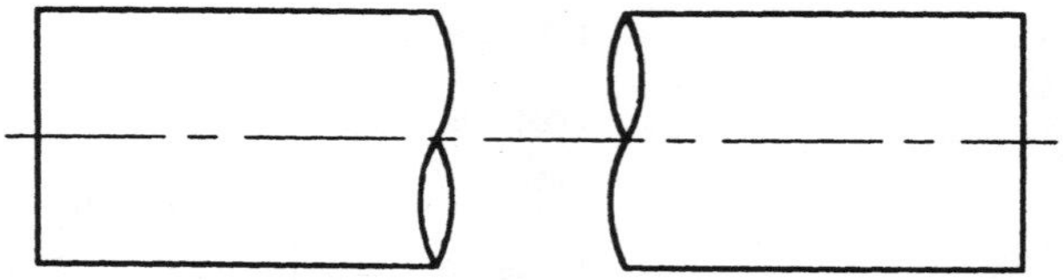

Fig 8.10

10

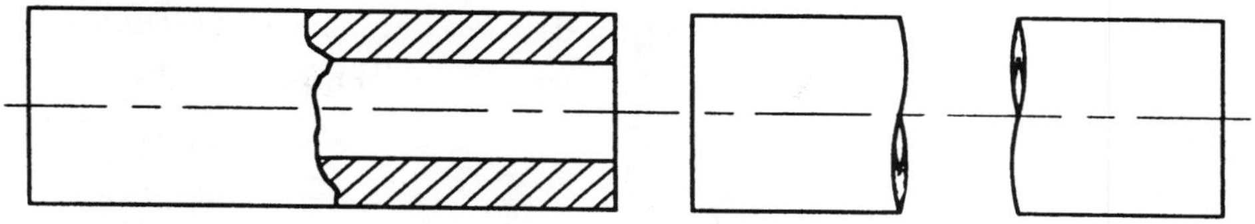

Fig 8.11

11 Welding is a method of permanently joining two pieces of metal.

12 Four (butt, lap, T- and corner joints).

13 A symbol that specifies the type of weld that is to be used.

14 The weld goes all around the join, i.e. the weld run starts and stops at the same point.

15

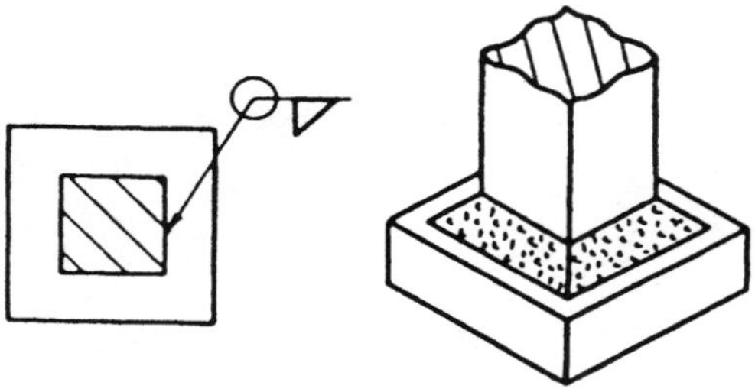

Fig 8.22

16 The weld is not done in the workshop but in the field.

17

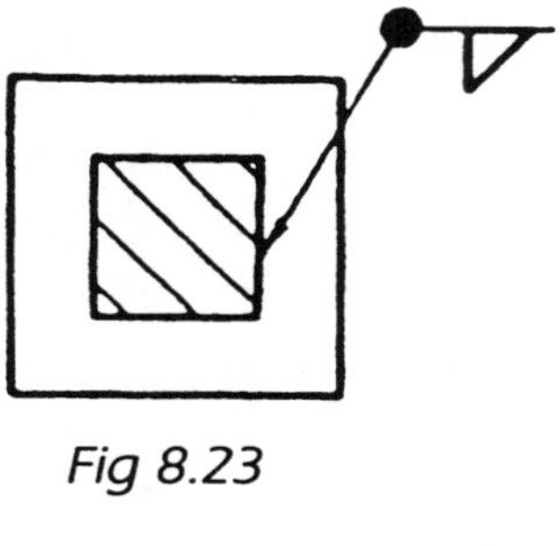

Fig 8.23

18

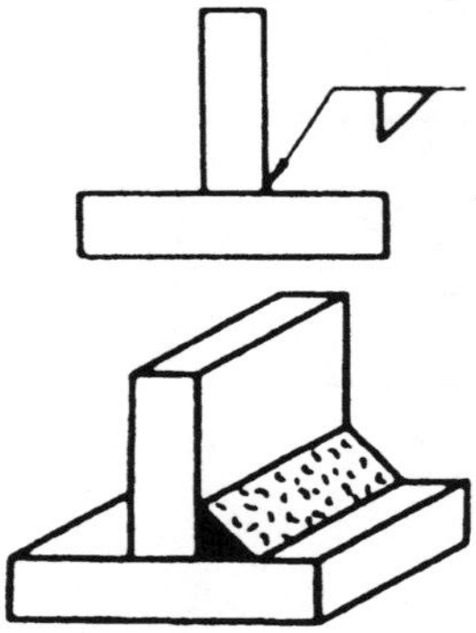

Fig 8.16

19

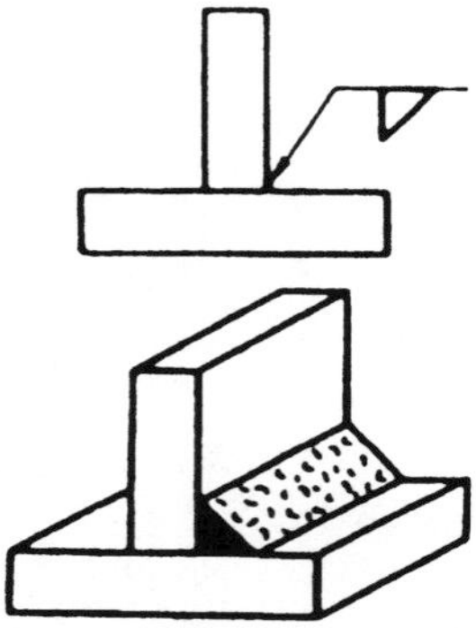

Fig 8.16

20

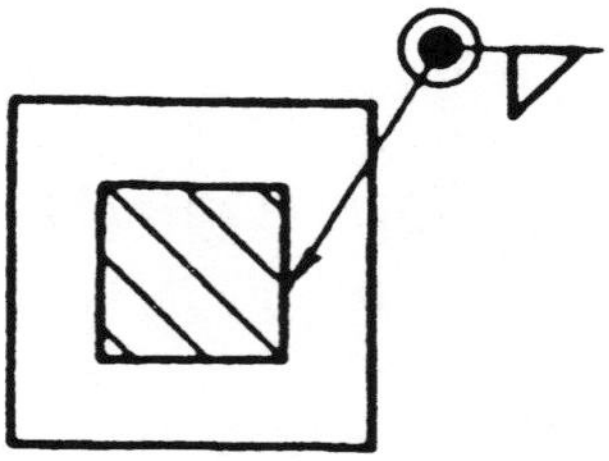

Fig 8.24

Unit 9

Fasteners

Study objectives

After studying this unit, you should be able to identify and make freehand sketches of the following components:

- ◇ Threaded fasteners
- ◇ Locking devices
- ◇ Riveted joints.

1 Introduction

In unit 8 you learned how to represent threaded fasteners in a drawing. In this unit you will learn about the different types of threaded fasteners.

Threaded fasteners are used for holding together parts which may require dismantling. They are identified by the shape of their heads.

These are all methods or joining parts to make an assembly. You must be familiar with these methods when making assembly drawings in the next unit.

2 Types of threaded fasteners

Threaded fasteners are commonly known as bolts or screws. Recall that screws are threaded all the way and bolts have an unthreaded portion called the shank.

Both bolts and screws are identified by the shape of their heads. Different applications require different types of head.

2.1 Bolt heads

Table 9.1

NAME	SKETCH	REMARKS
Hexagonal head bolts	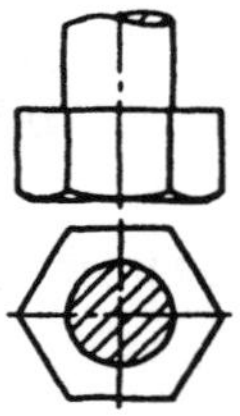 Fig 9.1	These are the most common type used.
Square head bolts	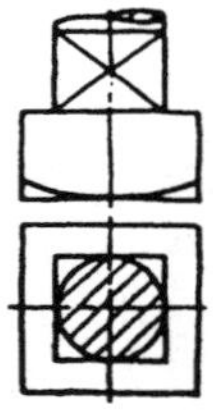 Fig 9.2	These are used where the bolt head has to be recessed to leave the face flush.
T-head bolts	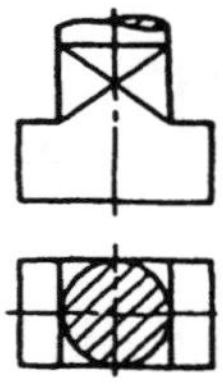 Fig 9.3	These are used for bolting jobs to slotted machine tables.
Cheese head bolts	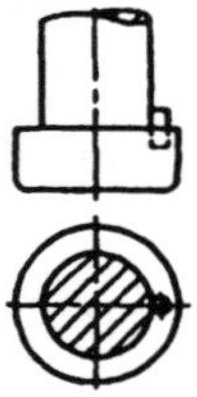 Fig 9.4	These have a "snug" under the head to prevent rotation. A snug is a projection piece forged under the head of a bolt.

NAME	SKETCH	REMARKS
Hook bolts	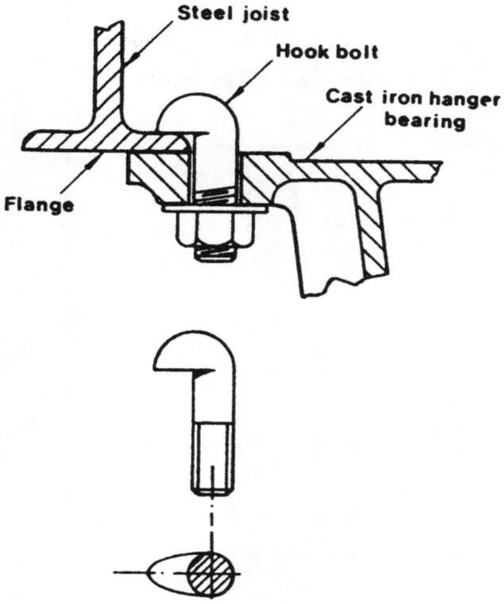 Fig 9.5	These are used to clip plate edges.

2.2 Screw heads

Table 9.2

NAME	SKETCH	REMARKS
Round head screw	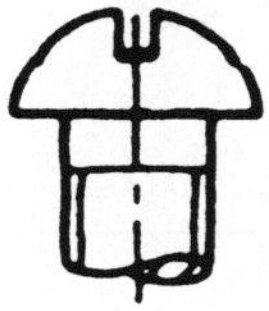 Fig 9.6	
Cheese head screw	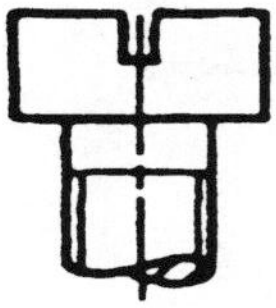 Fig 9.7	

NAME	SKETCH	REMARKS
Fillister head screw	 Fig 9.8	
Socket head screw	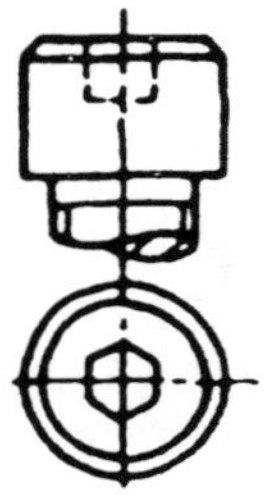 Fig 9.9	An Allen key is used and not a screwdriver.
Countersunk head screw	 Fig 9.10	This is used in a countersunk hole to leave the face flush.
Instrument head screw	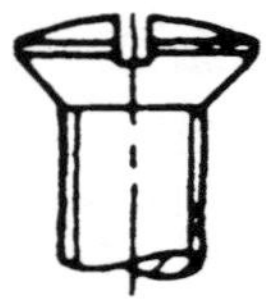 Fig 9.11	This is used by instrument makers and electricians.

3 Locking devices

Nuts and bolts which are subjected to vibrations are likely to work loose and come off. To prevent this, locking devices are used. Some of the locking devices are nuts with a certain device built into them, as you will see in the sketches below. Some use other devices such as washers and pins to lock.

You must be able to identify these locking devices and sketch them freehand.

Spring washer (most commonly used)

When the nut is screwed down, the spring washer keeps it tight. Spring washers are used on **pedestal bearings** and for **securing motor car radiators**, among other applications.

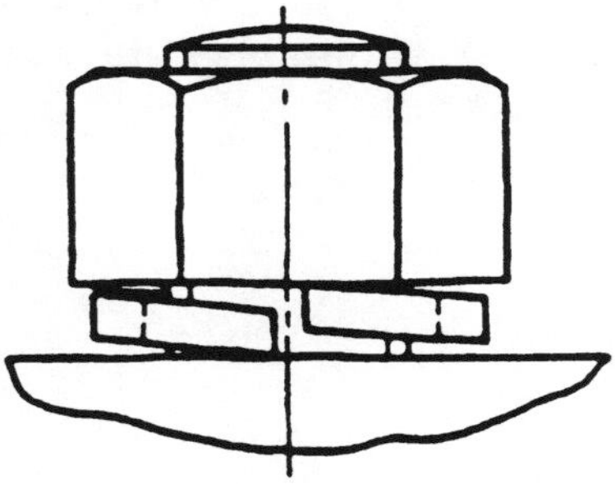

Fig 9.12

Simmond's nut

As the nut is screwed on, the bolt cuts its own thread on the inside of the fibre or plastic collar which is held in a 'housing' at the top of the nut. This type of nut is used on **aircraft**.

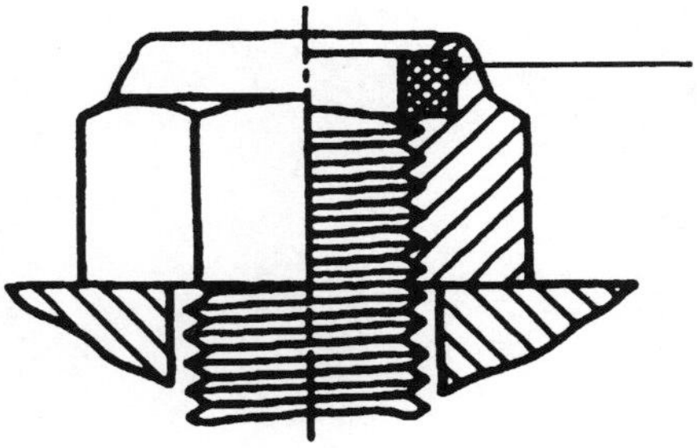

Fig 9.13

Lock nut

The thin lock nut is screwed down tightly and held with a spanner while the other nut is screwed on and tightened. This method is used to **set the gaps of the tappets for the valve clearances on a motor car**.

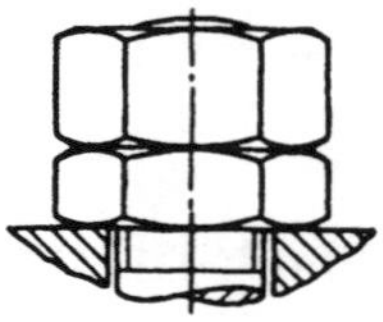

Fig 9.14

Slotted nut

Slotted nuts are ordinary hexagonal nuts slotted at the top to take a split pin or a pin which passes through a hole in the bolt. Slotted nuts are usually used on the front or rear **wheels of a motor car**.

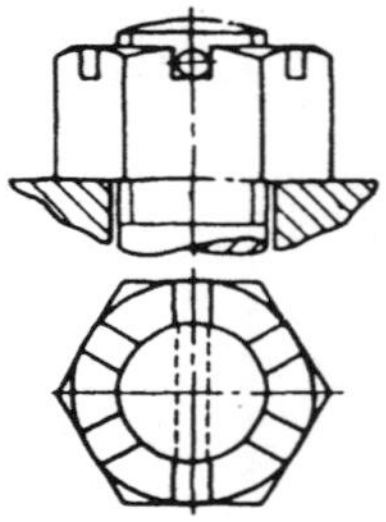

Fig 9.15

Castle nut

This type has a slotted built-on ring at the top of the hexagonal part to accommodate a pin or a split pin. Also widely used in **motor car construction** and on **aircraft**.

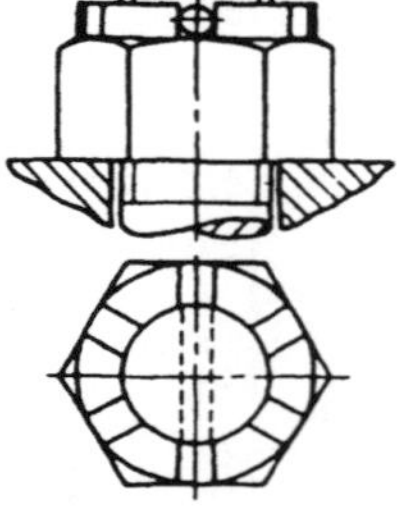

Fig 9.16

Tab washer

This type of washer is similar to an ordinary washer but has two tabs, one of which is bent upwards against the nut and the other, downwards against the side of the component. Tab washers are used mostly in **connecting rods** of petrol engines.

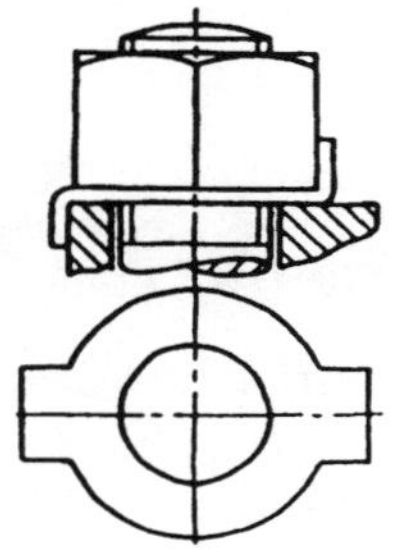

Fig 9.17

4 Riveted joints

Unlike removable fasteners such as bolts, nuts and screws, rivets are used to make **permanent joints** which are mainly used for **boiler seams and structural steelwork**.

Snap-head rivet

This type of rivet is most commonly used in engineering for a permanent joint.

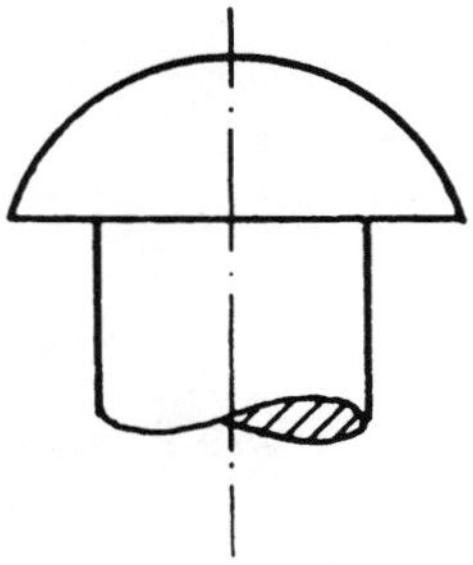

Fig 9.18

Pan-head rivet

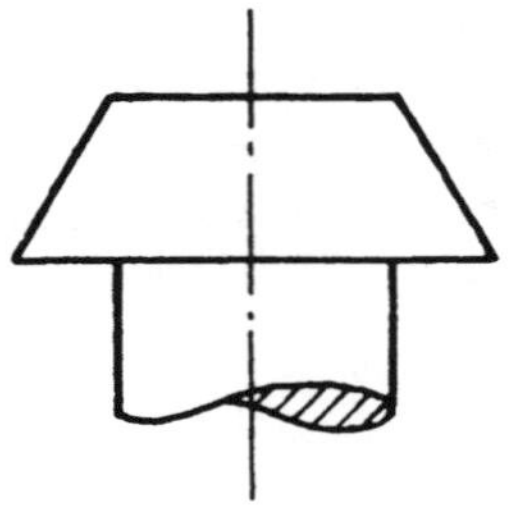

Fig 9.19

Countersunk rivet

This is used when the face of the plate must remain flush/level.

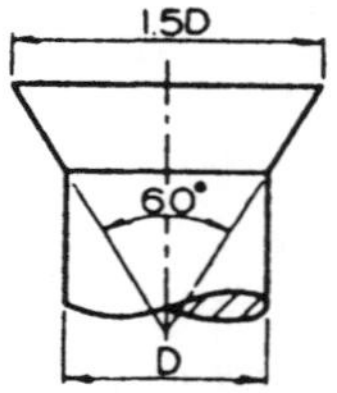

Fig 9.20

4.1 Single or double rivet joints

Single or double refers to the number of rows of rivets running through each of the two plates being joined. In a **lap** joint, the edges of the two plates to be joined overlap. In a **butt** joint, the joining edges butt together and are covered by one or two cover plates, also called straps.

Lap joint

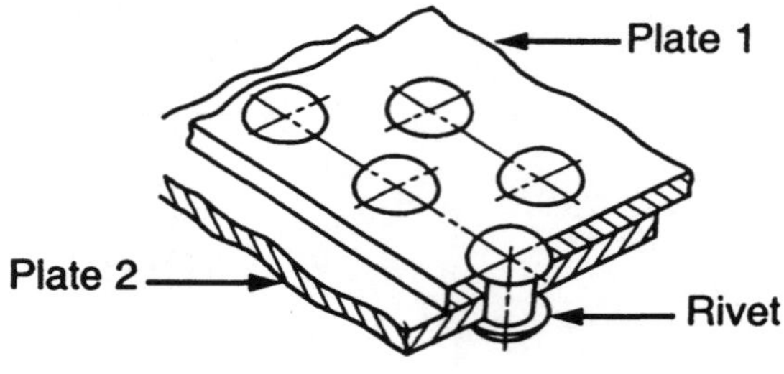

Fig 9.21

Butt joint with a cover strip

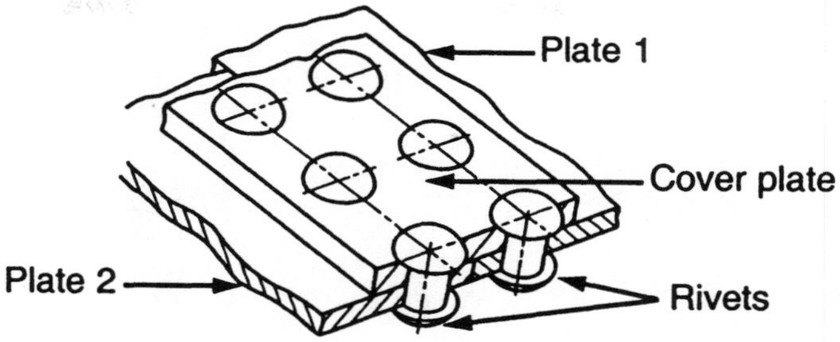

Fig 9.22

Single-riveted lap joint

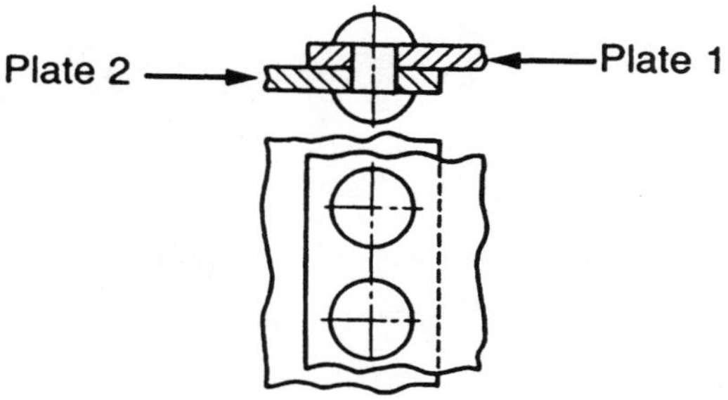

Fig 9.23

Double-riveted lap joint (zig-zag riveting)

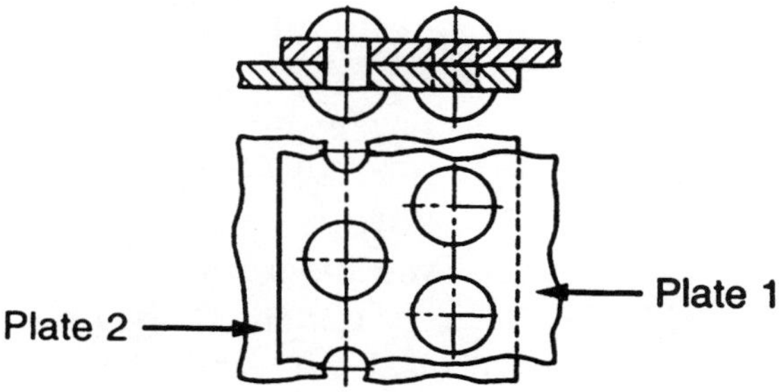

Fig 9.24

Double-riveted joint (chain riveting)

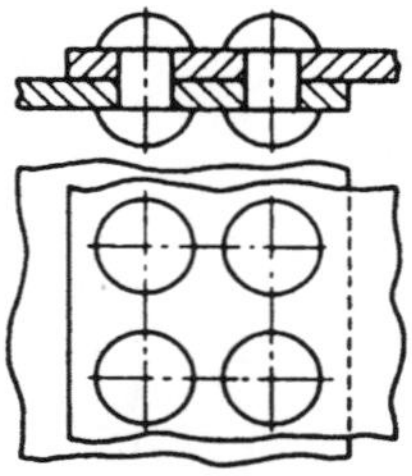

Fig 9.25

Single-riveted butt joint with a single cover plate (chain-riveted)

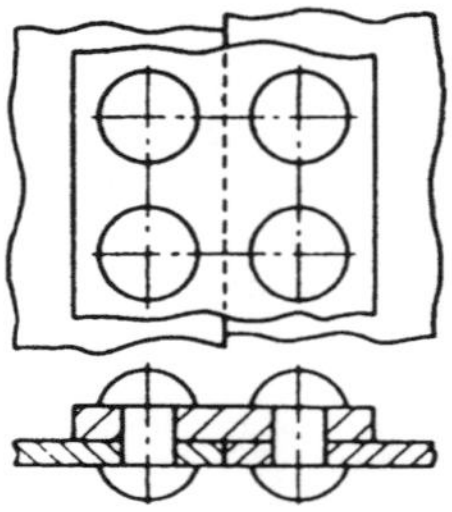

Fig 9.26

Double-riveted butt joint with a double cover plate (zigzag-riveted)

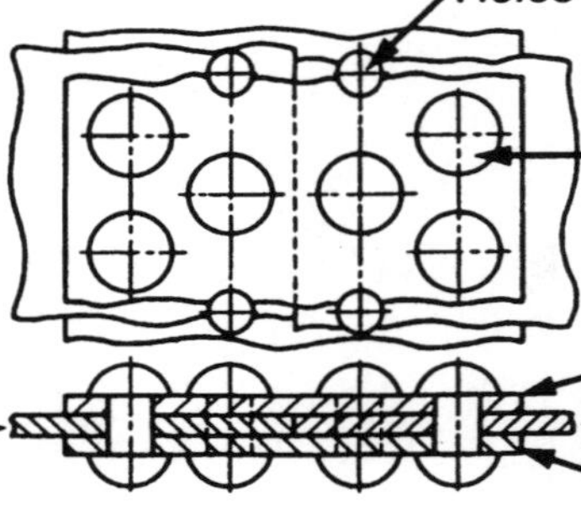

Fig 9.27

5 Summary

In this unit you have learnt the different types of fasteners and locking devices. You should be able to identify and make freehand sketches of all these components.

Self-evaluation

Question 1

Construct a freehand sketch, in good proportion, of two views of any three types of bolt heads shown in this unit.

Insert the titles.

Question 2

Make neat freehand sketches, in good proportion, of any three types of screw heads shown in this unit.

Question 3

Draw neat freehand sketches of each of the locking devices. Show an outside top view where necessary. Print the titles and ensure that the sketches are in good proportion.

Question 4

Which locking devices would you use on each of the following?

- **a** Wheel axle of a motor car
- **b** Fixing bolt of a pedestal bearing
- **c** Petrol-engine connecting rod
- **d** Bolts for securing motor car radiator
- **e** Aircraft construction

Question 5

Sketch freehand and in good proportion two **orthographic views** of the riveted joints shown in this unit. Add the titles.

Question 6

Sketch freehand and in good proportion the following types of welded joints. Insert the welding symbols and the titles.

- **a** Butt joint
- **b** Lap joint
- **c** T-joint
- **d** Fillet weld

Also sketch these welded joints isometrically.

Answers

All the questions in this evaluation, except question 4, tested your ability to apply the techniques you learnt in all the previous units. In an examination, the examiner would look at the following five aspects. The weighting of each aspect is also given. The same weighting can be applied to exercise 1 as well.

1	Good proportion according to given measurements	50%
2	Linework (e.g. same thickness)	25%
3	Centre lines	10%
4	Accuracy	10%
5	Neatness	5%

Answers to question 4:

- **a** Slotted nut
- **b** Spring washer
- **c** Tab washer
- **d** Spring washer
- **e** Simmond's nut

Unit 10

Assembly drawings

Study objectives

After studying this unit, you should be able to:

- ◇ Apply the correct layout of an assembly drawing
- ◇ Draw sectional and assembly drawings, applying the hatching lines properly
- ◇ Distinguish between detail drawings and assembly drawings
- ◇ Insert leader lines to identify parts
- ◇ Make a parts list
- ◇ Draw a complete assembly drawing.

1 Introduction

In unit 7 we discussed the different types of sections that are applied in detailed sectional drawings. We also discussed cutting planes as well as hatching in some detail.

In this unit we are going to explain, and show you how to draw, assembly drawings together with sections, item numbers, parts, lists, title and scale, and the projection symbol. Assembly drawings are drawings where you place the different parts together to form an 'assembly of parts'. The engine of a motor car is a good example. It consists of pistons, pulleys, bearings, etc. that are put together (assembled) to form a unit. The **assembly drawing** is used mainly to:

- ◇ Represent the proper relationship of the mating parts of a machine and their different functions.
- ◇ Give a general idea of what the finished assembly should look like.
- ◇ Distinguish the different parts with the aid of item numbers.

2 Hatching sectional drawings

To show the different parts on an assembly drawing clearly you use **hatched lines**. To distinguish between the different parts, the space between the hatched lines varies, and the angle of the direction of these lines also changes.

Note the three parts, two of which have been sectioned

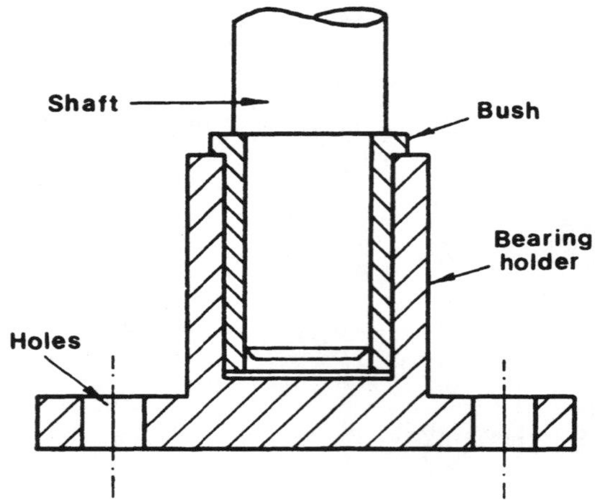

Fig 10.1 Sectional assembly drawing of a shaft, bush and bearing holder

The figure above shows a shaft, bush and bearing holder fitted together to form an assembly. The hatched lines are at different angles for each adjacent part.

If no sectioning of the different parts is done we call such a drawing an **outside assembly drawing**. When the parts are sectioned, we refer to the drawing as a **sectional assembly drawing**.

3 Sections of sectional drawings

Example 1 Fig 10.2 shows two outside views in first angle orthographic projection of a bracket assembly. Draw in a first-angle orthographic projection and full scale the following views:

1. A sectional front view on cutting plane A-A
2. A sectional left view on cutting plane C-C

3 A sectional top view on cutting plane B-B.

Item numbers are given to show the different parts clearly.

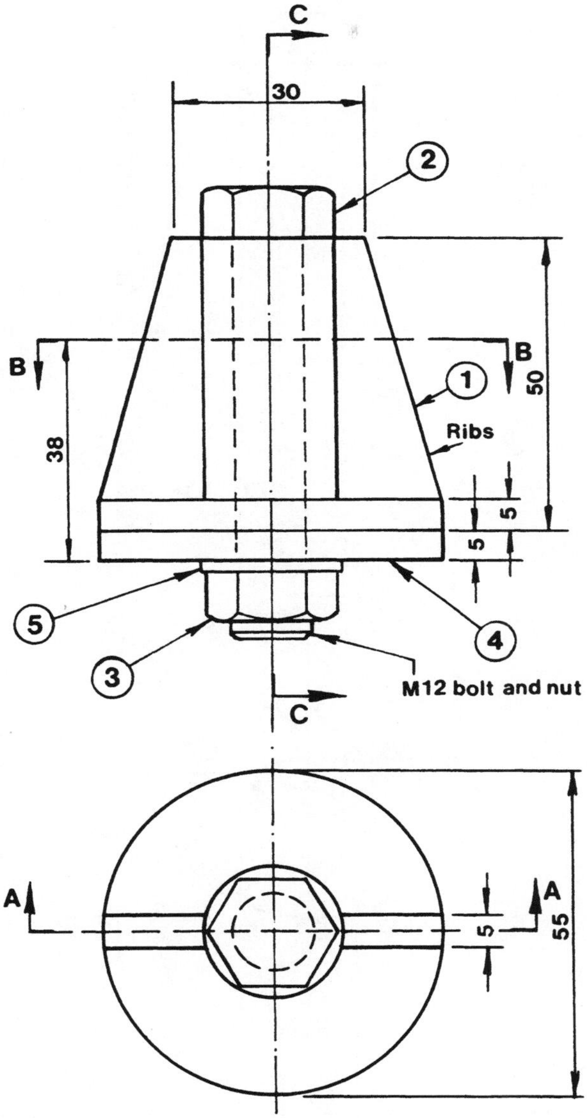

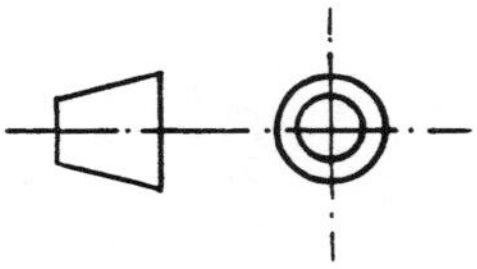

Fig 10.2 Outside bracket assembly

Answer

Remember that we do not use hatched lines for bolts, nuts or flanges/ribs

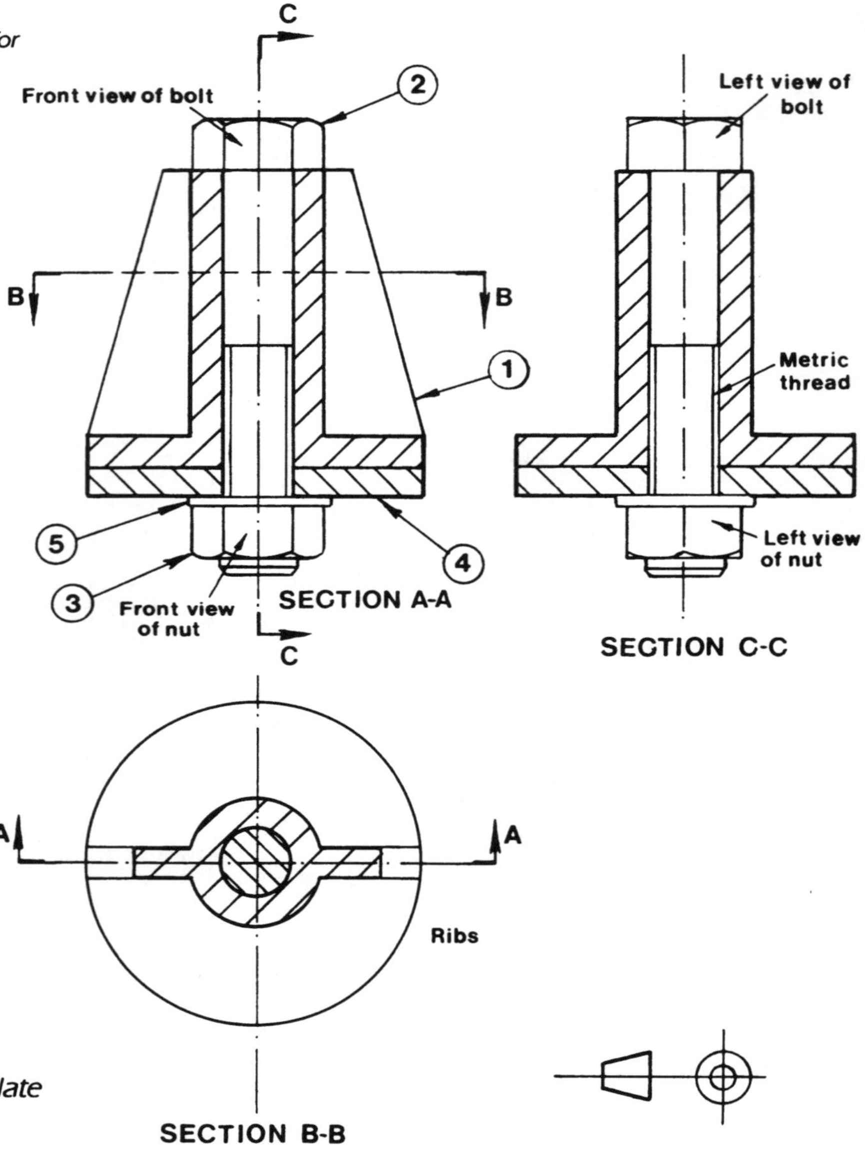

1 Body
2 Bolt
3 Nut
4 Bottom Plate
5 Washer

Fig 10.3

Activity 1

Activity 1 has two exercises in which you have to draw a few views of each of the two drawings which you have to interpret. Once you have completed the exercises in a rough-work book, you may look at the answers section at the end of the unit to check whether your interpretation and drawings are correct.

1 Fig 10.4 shows two views of a V-pulley assembly. Draw the following views in a third-angle orthographic projection to a scale of 1 : 1:

a Full sectional front view

b A right view

c A top view.

Do not show any hidden detail.

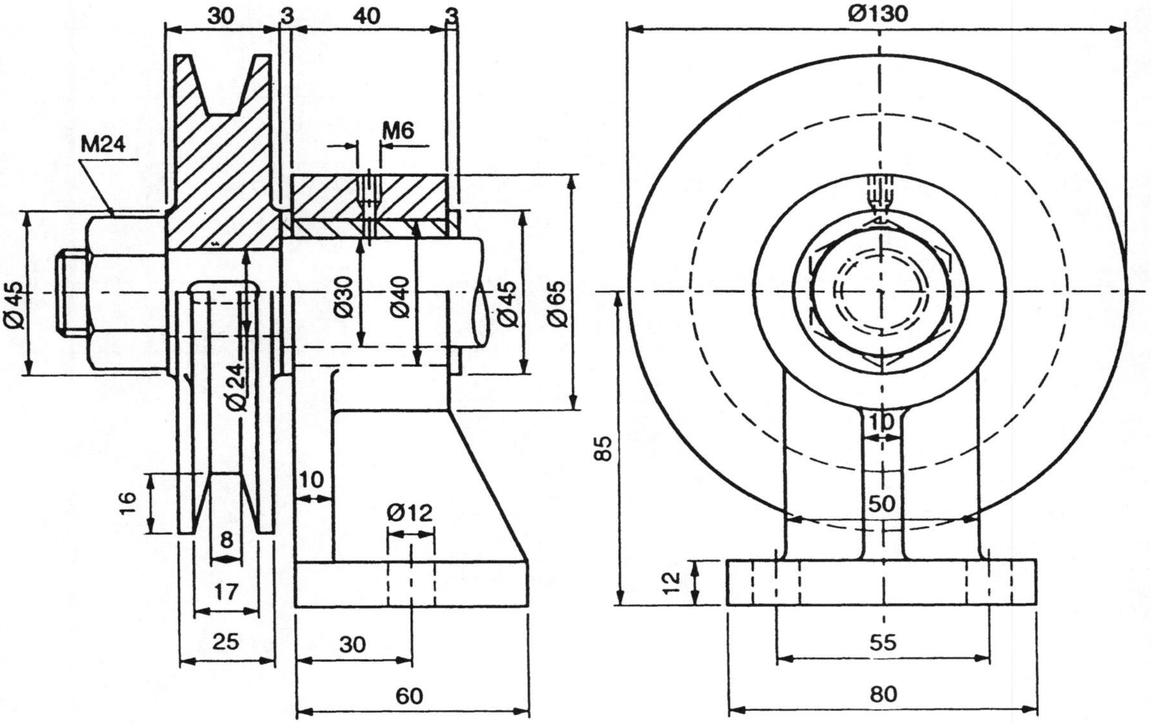

Fig 10.4

Remember to identify the different parts and concentrate on the correct hatching.

2 Fig 10.5 shows a flange-coupling in section. To make the assembly more accurate, the one side fits perfectly into the other one. Draw, according to third-angle orthographic projection, the following views to a scale of 1:2.

a **A half-sectional front view with the top half in section.**

Do not show any hidden detail.

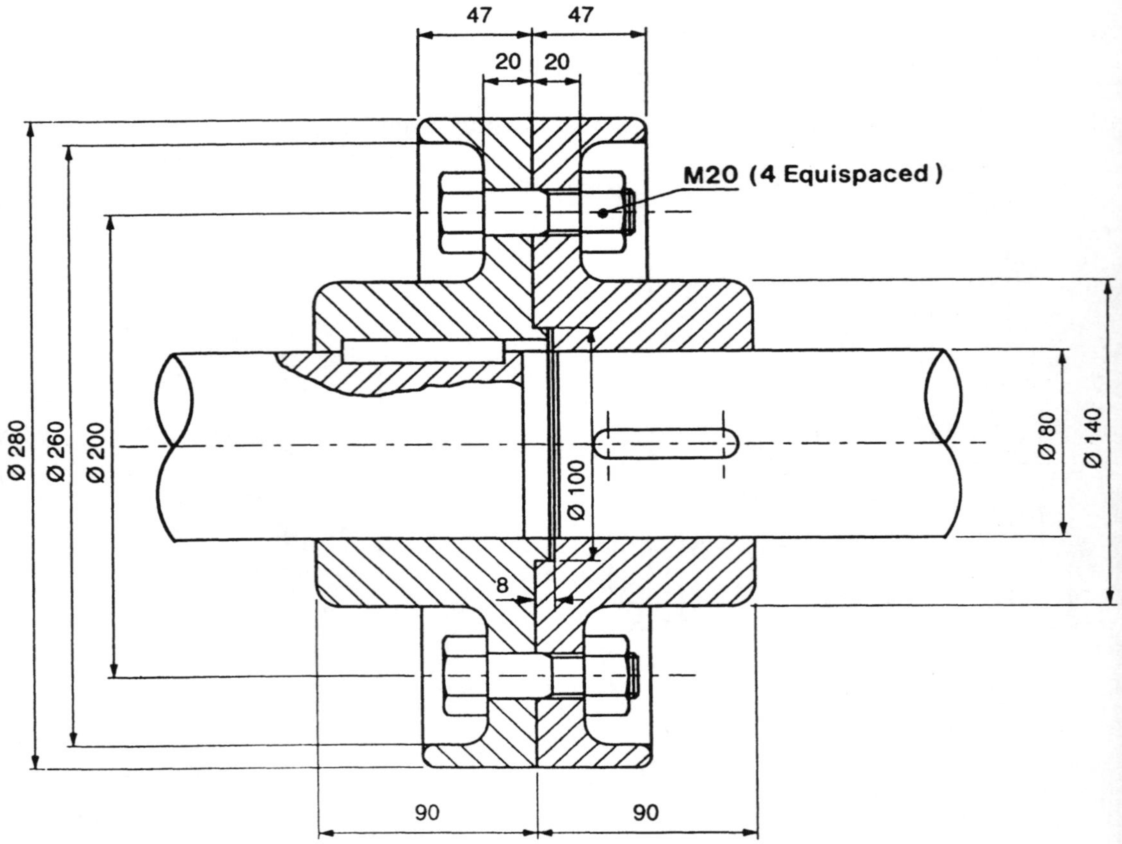

Fig 10.5 A flange-coupling in section

A flange-coupling is used to join two shafts.

4 A typical assembly drawing

Fig 10.6 shows a typical example of a complete crane hook assembly. The finished drawing must have all the items or parts numbered and numbered in the correct position. The assembly drawing must also include:

- ◇ A parts list with all the parts listed.
- ◇ The material for the parts must be listed.
- ◇ The number per assembly must also be stated.

◇ **The title and scale and also the projection symbol must be shown.**

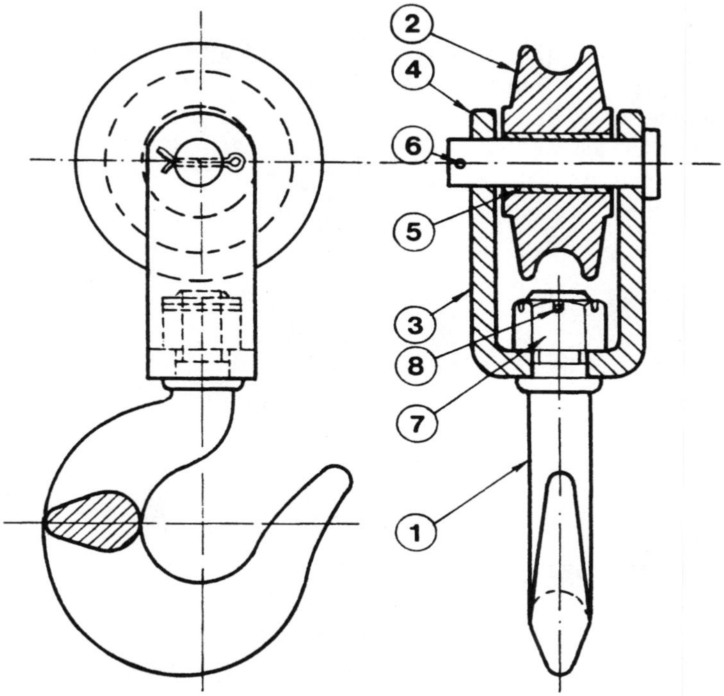

CRANE HOOK SCALE 1:1

PART NO	NAME OF PART	MATL NR	OFF
1	HOOK	M.S.	1
2	PULLEY	M.S.	1
3	BRACKET	M.S.	1
4	SHAFT PIN	M.S.	1
5	BUSH	BRONZE	1
6	COTTER PIN	M.S.	1
7	HEX NUT	M.S.	1
8	LOCKING PIN	M.S.	1

Fig 10.6

5 Item numbers

Leader lines ending in an arrow or a dot are used to show the different parts or items as shown in fig 10.7a. Fig 10.7b shows the proper dimensions for drawing item numbers.

When the leader lines end in a dot they are usually used inside the drawing. Leader lines ending with an arrow are used on the outside lines of the parts.

Each part is identified by a ringed number placed at the end of a leader line.

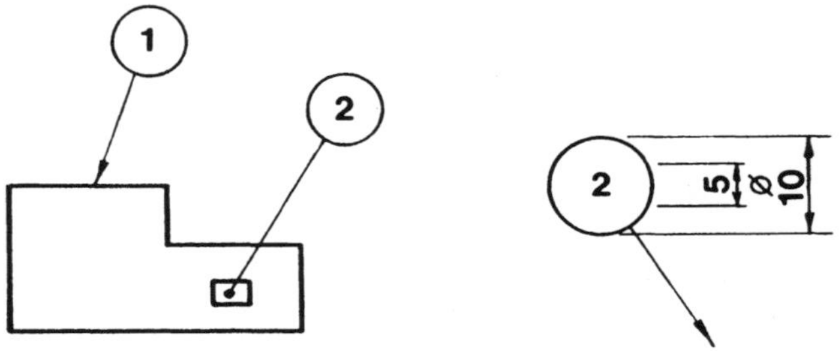

a *b*

Fig 10.7a–b

6 Parts lists

A parts list is an itemized list of the different parts of a design, showing the part number, name of the part, material and quantity required. The various parts are numbered in order of importance. Fig 10.8 shows an example of a parts list and the dimensions of the block work.

Standard parts such as bolts, screws, nuts, washers and splitpins should also be listed in the parts list.

	PART NO	NAME OF PART	MATL NR	OFF
7				
7	1	BODY	C.I.	1
5	2	PLATE	M.S.	1
5	3	HEX. NUT	M.S.	1
5	4	STUD	M.S.	1

Fig 10.8

Use your own discretion for these distances.

7 How to start an assembly drawing

- ◇ Detailed parts are given separately (see fig 10.10).
- ◇ The item numbers of the detailed parts are also given. More than one view of the detailed parts is also given, as is the projection symbol.
- ◇ Fig 10.10 gives you all the separate detailed parts.
- ◇ Complete the assembly drawing with all the parts assembled in position (see fig 10.9).
- ◇ Insert the item numbers and complete the parts list as shown.

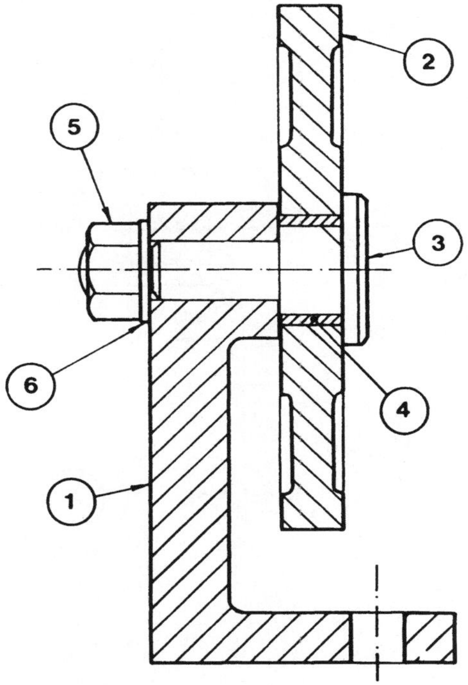

IDLER PULLEY SCALE 1 : 1

ITEM NO	DESCRIPTION	MATL NR	OFF
1	BODY	MILD ST.	1
2	PULLEY	MILD ST.	1
3	PIN SHAFT	MILD ST.	1
4	BUSH	BRASS	1
5	HEX.NUT	M.S.	1
6	WASHER	M.S.	1

Fig 10.9

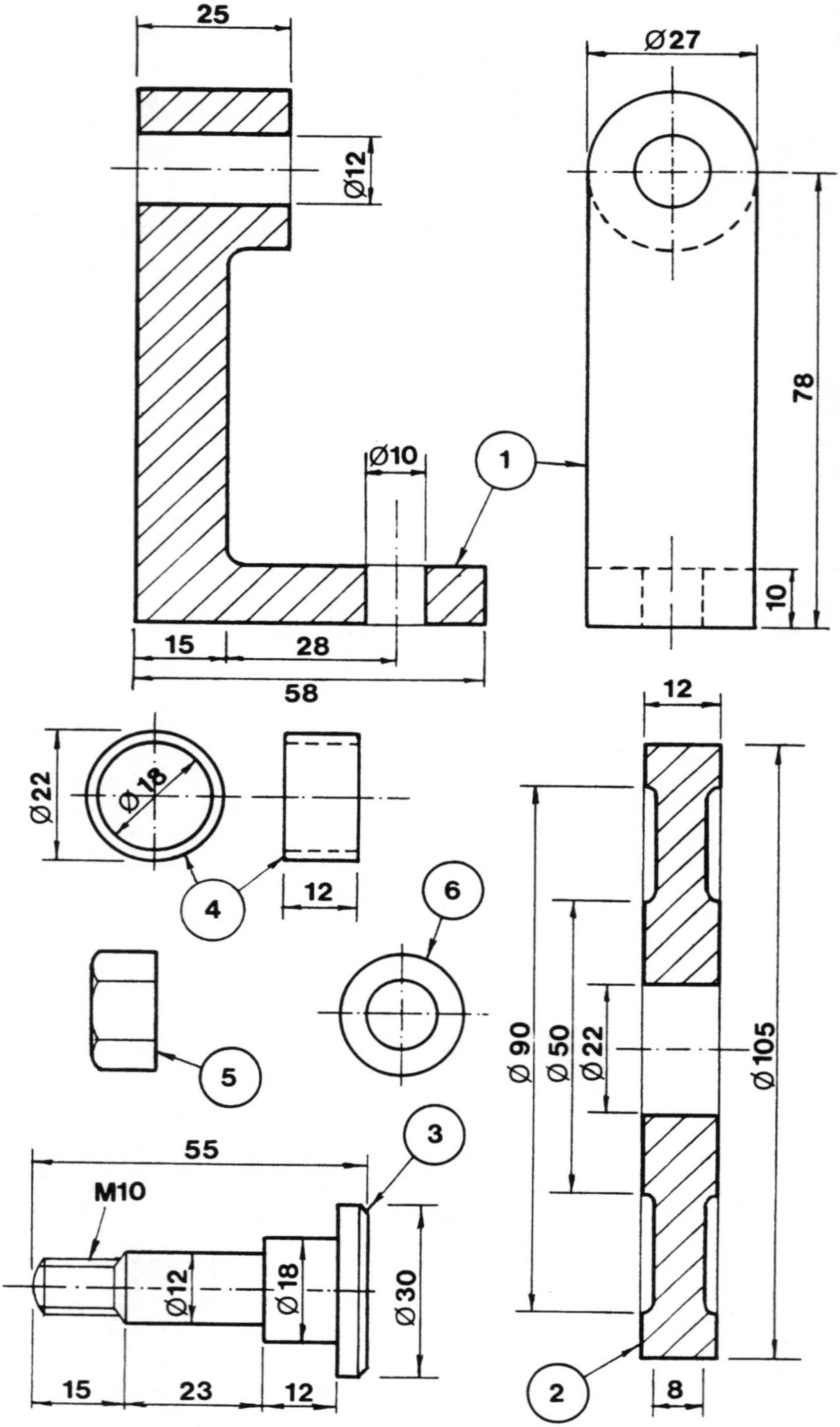
25
Ø27
Ø12
78
Ø10
1
10
15
28
58
12
Ø22
Ø18
4
12
6
5
Ø 90
Ø 50
Ø 22
Ø 105
55
3
M10
Ø12
Ø18
Ø30
15
23
12
2
8

Fig 10.10

Activity 2

Question 1

Detail drawings, in first angle orthographic projection, of a locating device together with a key to its assembly are shown in fig 10.11. The details are as follows:

- ◇ Body (cast iron) – one required – Item 1
- ◇ Plate (mild steel) – one required – Item 2
- ◇ Hexagonal nut (mild steel) – one required – Item 3
- ◇ Stud (mild steel) – one required – Item 4.

Do not copy the given views of the separate parts but draw full size (1 : 1) and in third-angle orthographic projection, the following views of the assembled parts. Remember that assembly drawings do not include dimensions.

- ◇ A sectional front view (full sectioned).
- ◇ An outside right view.
- ◇ An outside top view.
- ◇ Insert a parts list and also the item numbers.
- ◇ Insert the title and the scale together with the projection symbol.

Question 2

Details of a swivel and base are shown in fig 10.12. The headed pin passes through the base so that the head rests in the counterbore. The swivel is then put on the end of the pin and secured by the cotter. The details are as follows:

- ◇ Base (cast iron) – one required – Item 1
- ◇ Swivel (cast iron) – one required – Item 2
- ◇ Headed pin – (mild steel) – one required - Item 3
- ◇ Cotter – (mild steel) – one required – Item 4.

Do not copy these given views of the separate parts but draw full size (1 : 1) and in first-angle orthographic projection the following views of the assembled parts:

1. Outside front view.
2. Outside left view.
3. Full sectional top view.
4. Draw up a parts list and also insert the item numbers.
5. Insert the title and scale and also the projection symbol.

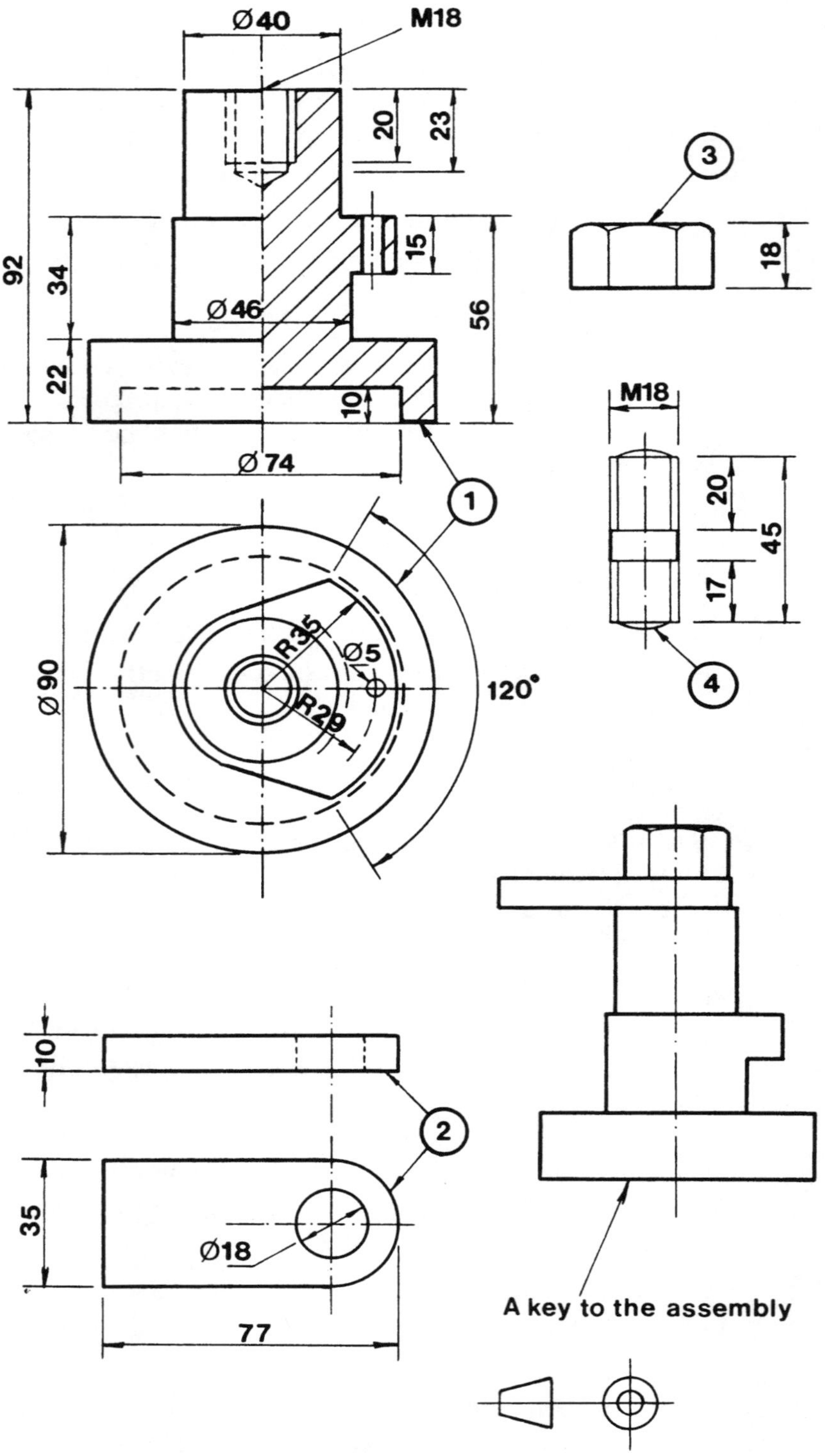
Ø40
M18
20
23
92
34
15
Ø46
56
22
10
Ø74
1
3
18
M18
20
45
17
4
Ø90
R35
Ø5
R29
120°
10
2
35
Ø18
77
A key to the assembly

Fig 10.11

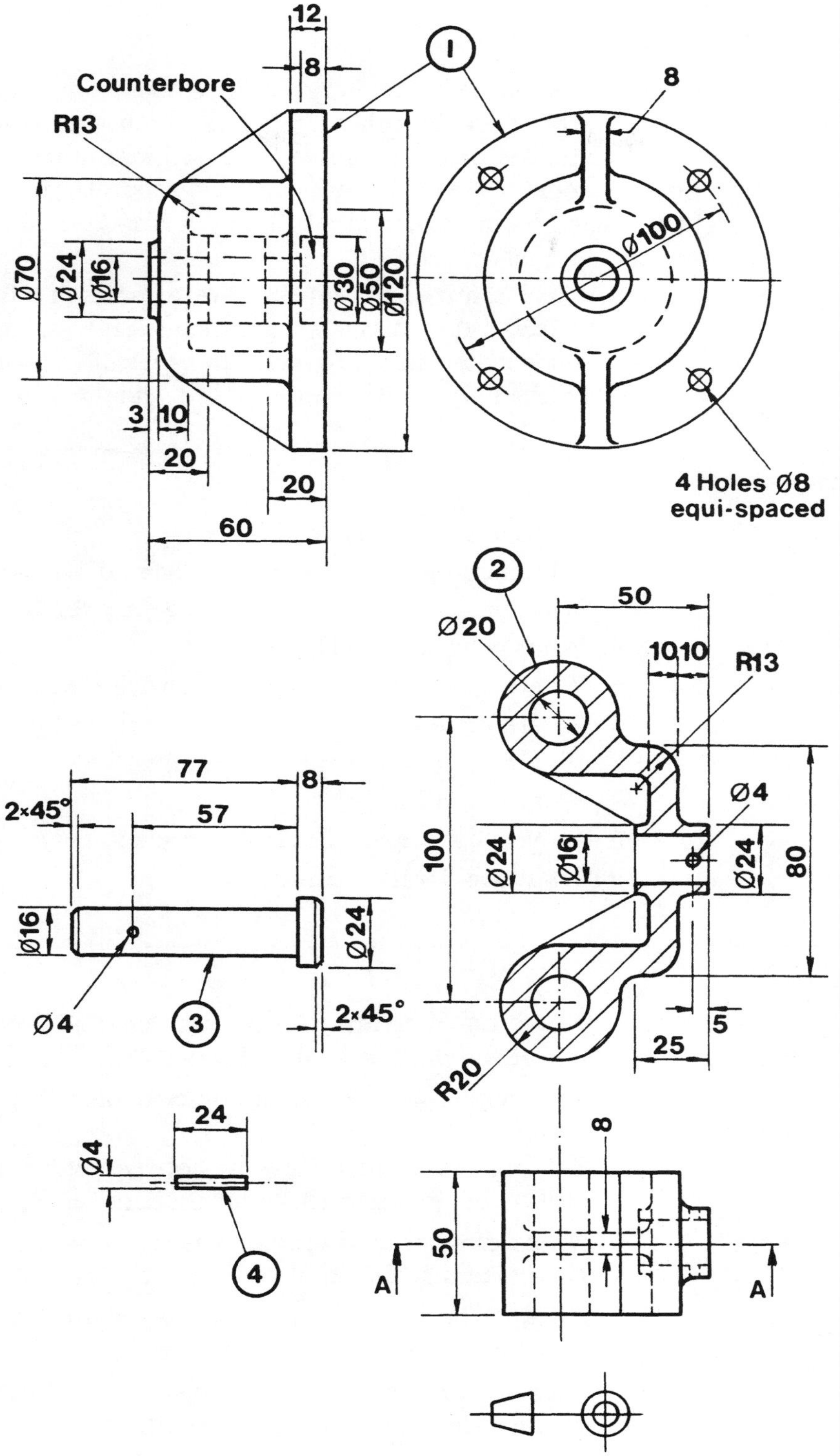
12
8
Counterbore
R13
1
8
Ø70
Ø24
Ø16
Ø30
Ø50
Ø120
Ø100
3
10
20
20
60
4 Holes Ø8
equi-spaced
2
50
Ø20
10
10
R13
77
8
2×45°
57
Ø4
100
Ø24
Ø16
Ø24
80
Ø16
Ø24
Ø4
3
2×45°
5
25
R20
24
Ø4
4
8
50
A
A

Fig 10.12

8 Summary

Detail drawings serve as working drawings from which the individual parts of an assembly are made. Each part is drawn separately, complete with the dimensions and manufacturing instructions. Assembly drawings show how the separate parts of a design fit together. Each part is identified by a ringed number placed at the end of a leader line.

Parts lists are itemized lists of the several parts of a design showing the part number, name of part, material and quantity required. The various parts are listed in order of importance. Standard parts like bolts, nuts, washers, etc. should also be listed.

Self-evaluation

1. What are hatching lines?
2. Why must the angle vary for each adjacent part?
3. Why must the distance or space vary for hatched lines?
4. What is an assembly drawing?
5. Where do you use a leader line with a dot?
6. Where do you use a leader line with an arrow?
7. Why must you use an item number?
8. What is a parts list?
9. Do you list standard parts in a parts list?
10. What are detailed drawings?

Answers

1. Hatching lines are used to show the different sections on all the parts assembled together.
2. The angles vary for each adjacent part to prevent confusion.
3. For big parts the spaces in between the hatched lines will be bigger than those for smaller parts.
4. An assembly drawing is a drawing in which you put all the parts together to form a functional unit.
5. A leader line ends in a dot and is usually used inside an assembly drawing to show the part.
6. A leader line ending in an arrow is used at the outline of the assembly drawing or detail.
7. Item numbers are used for the different parts in an assembly drawing.

8 Part lists are itemized lists of the different parts used in an assembly drawing, showing the part or item number, name of part, material and quantity required.

9 Standard parts should also be listed in a parts list.

10 Detail drawings serve as working drawings from which the individual parts of an assembly are made.

In this case you should have your answers to the activities evaluated by fellow students or your lecturer.

Solution to Activity 1:

Fig 10.13 will help you to see whether you have interpreted the instruction correctly. The details of the answer should be verified with either fellow students, tutors or lecturers.

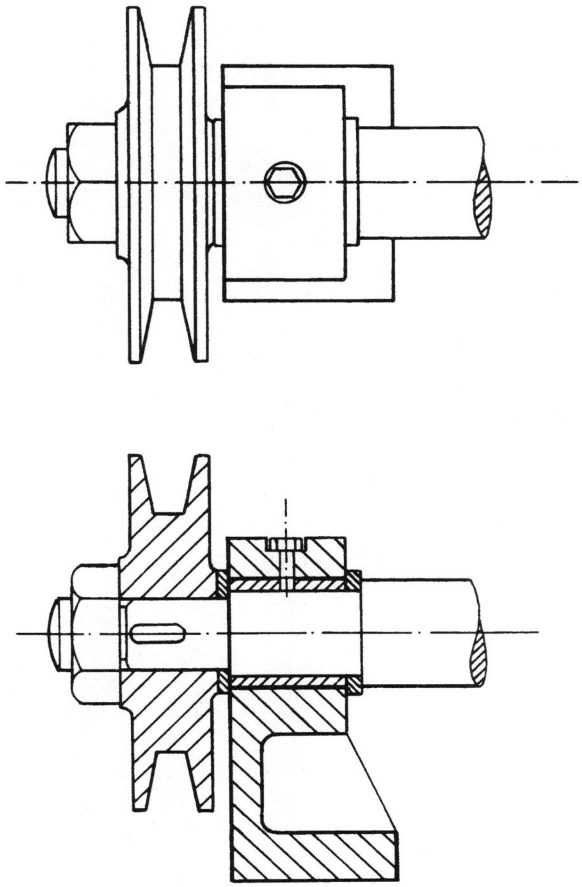

Fig 10.13

Fig 10.14 shows how the drawing should be interpreted.

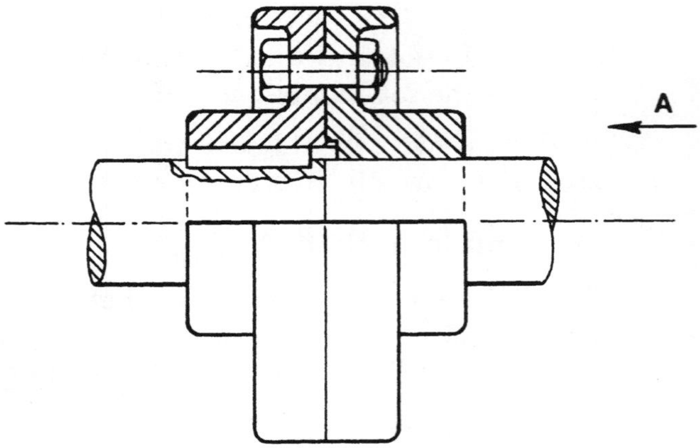

Fig 10.14

Pipe drawings (chemical)

Study objectives

After studying this unit, you should be able to:

- ◇ Distinguish between the different kinds of piping
- ◇ Distinguish between the different pipe joints and fittings
- ◇ Distinguish between the two kinds of pipe drawings
- ◇ Insert the different pipe drawing symbols correctly.

1 Introduction

Until recently, water was the only important fluid conveyed from place to place through pipes. Today nearly every conceivable fluid is handled in pipes during its production, processing, transportation or utilisation. Fluids like liquid metals, sodium and nitrogen have been added to the list of more common fluids such as oil, water and acids which are transported through pipes. Many gases are also stored and delivered through piping systems.

Pipes are also used for hydraulic and pneumatic mechanisms and are used extensively for the controls of machinery and other equipment.

Pipe drawings are especially important to chemical engineers because of the extensive use of pipes in chemical plants.

In this unit we will be examining the different types of pipes, joints and fittings so that you can apply these principles to your drawings where relevant.

2 Kinds of piping

Steel and wrought iron pipes

These pipes convey water, steam, oil and gas and are commonly used under high temperatures and pressures.

Cast iron pipes

These pipes are often installed underground to carry water, gas and sewage.

Seamless brass or copper pipes

These pipes are used extensively in plumbing because of their ability to withstand corrosion.

Copper tubing

This kind of pipe is used in automotive, hydraulic and pneumatic design where vibration and misalignment are factors.

Plastic pipes

These pipes are usually used in a chemical plant due to their resistence to corrosion and chemicals. They are also easy to install. They are not recommended where heat and pressure are involved.

3 Pipe joints and fittings

Parts joined to pipes are called fittings. These fittings are used to change the size or the direction of the pipe branch.

There are generally three classes of fittings:

- ◇ Screwed – small diameter of pipes
- ◇ Welded – permanent fixtures
- ◇ Flanged – quick way to disassemble pipes.

Other methods such as soldering, brazing and gluing are used as well:

- ◇ Soldering – cast iron pipes
- ◇ Brazing – copper tubing
- ◇ Gluing – plastic pipes.

Pipe fittings are specified by the nominal pipe size, the name of the fitting and the material. The nominal pipe size will be given.

Some fittings, for example, tees, crosses and elbows, connect different sizes of pipes.

Screwed fittings

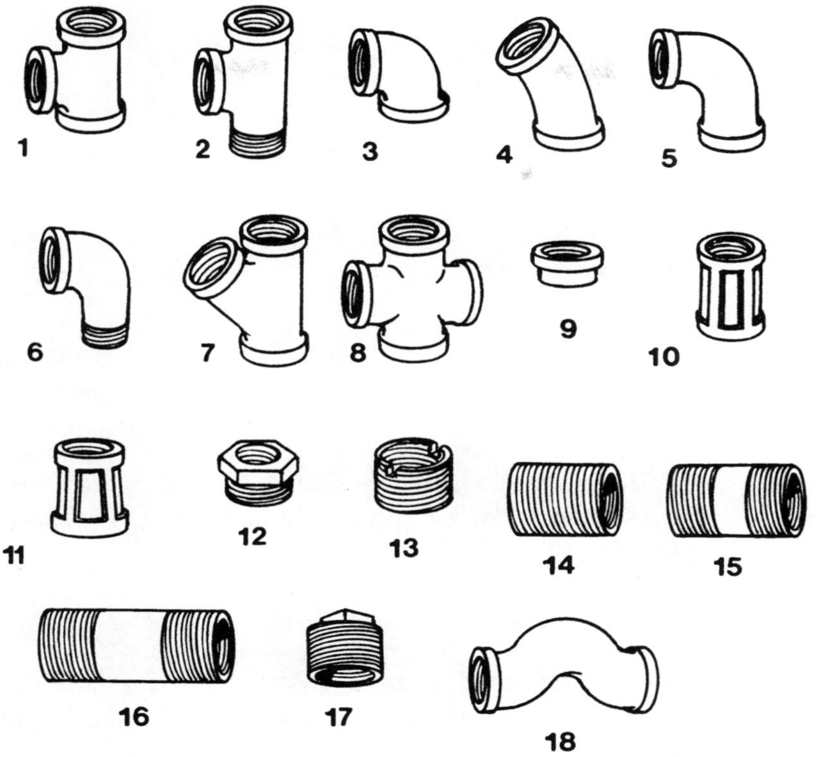

1 Tee
2 Service Tee
3 90° Elbow
4 45° Elbow
5 Reducing Elbow
6 90° Street Elbow
7 45° Y-Branch
8 Cross
9 Cap
10 R&L Coupling
11 Reducer
12 Hex Bushing
13 Bushing
14 Close Nipple
15 Short Nipple
16 Long Nipple
17 Plug
18 Bend

Flanged fittings

1 Tee
2 90° Elbow

Welded fittings

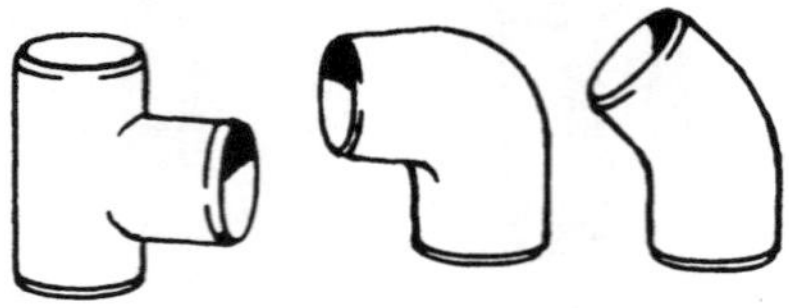

1 Tee
2 90° Elbow
3 45° Elbow

4 Pipe drawings

Pipe drawings show the size and location of pipes, fittings and valves and must be accurately described. There are two types of pipe drawings.

Double-Line Drawings

These types of drawings are suitable for catalogues and other applications where the appearance is of most importance.

They take longer to draw and are not recommended for production drawings.

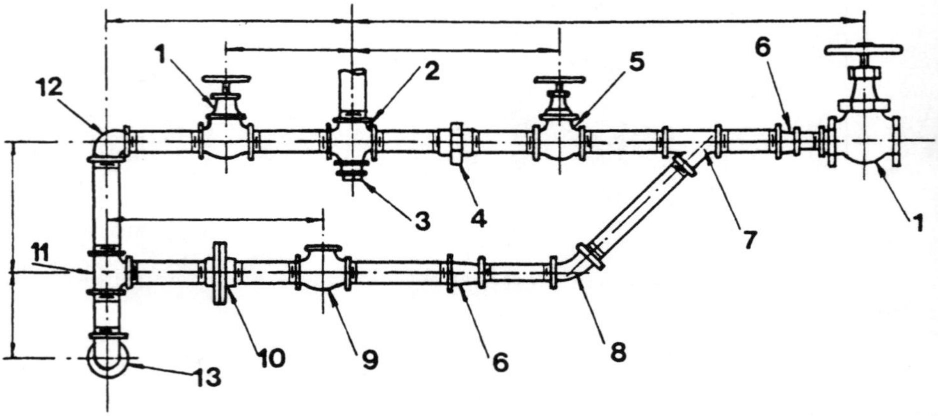

Fig 11.1

Listed fittings

1 Globe Valve **2** Cross
3 Plug **4** Union

5	Gate Valve	**6**	Reducer
7	Lateral	**8**	45° Elbow
9	Check Valve	**10**	Flanged Joint
11	Tee	**12**	Elbow
13	Elbow		

Single-Line Drawings

Single-line drawings of pipes use a single line to show the arrangements of the pipe fittings. They are much quicker to draw than double-line drawings and are therefore recommended for production drawings (see figs 11.2 and 11.3).

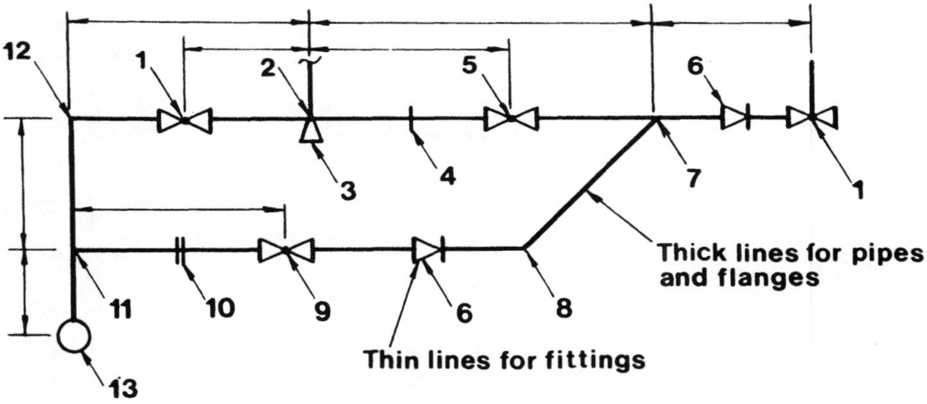

Fig. 11.2 Single-line drawing with new symbols

1	Globe Valve	**2**	Cross
3	Plug	**4**	Union
5	Gate Valve	**6**	Reducer
7	Lateral	**8**	45° Elbow
9	Check Valve	**10**	Flanged Joint
11	Tee	**12**	Elbow
13	Elbow		

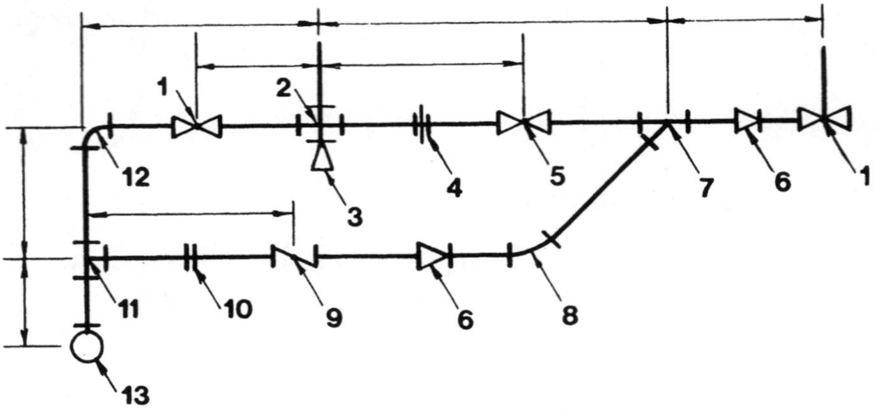

Fig. 11.3 Former single-line drawing symbols which are sometimes still used

5 Pipe drawing symbols

These symbols are generally used in chemical engineering, all are side view unless otherwise specified.

Table 11.1

	Flanged	*Screwed*	*Bell & Spigot*	*Welded*	*Soldered*
1 Bushing		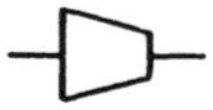	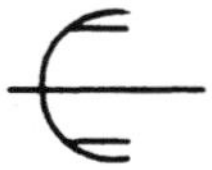		
2 Cap		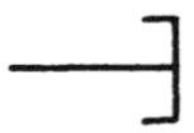			
3 Cross	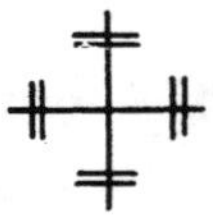		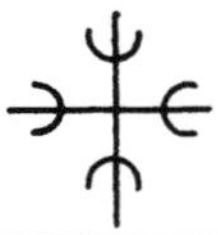		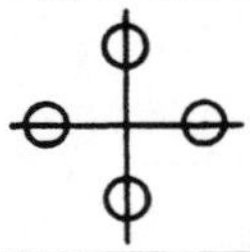
4 Elbow 45°					

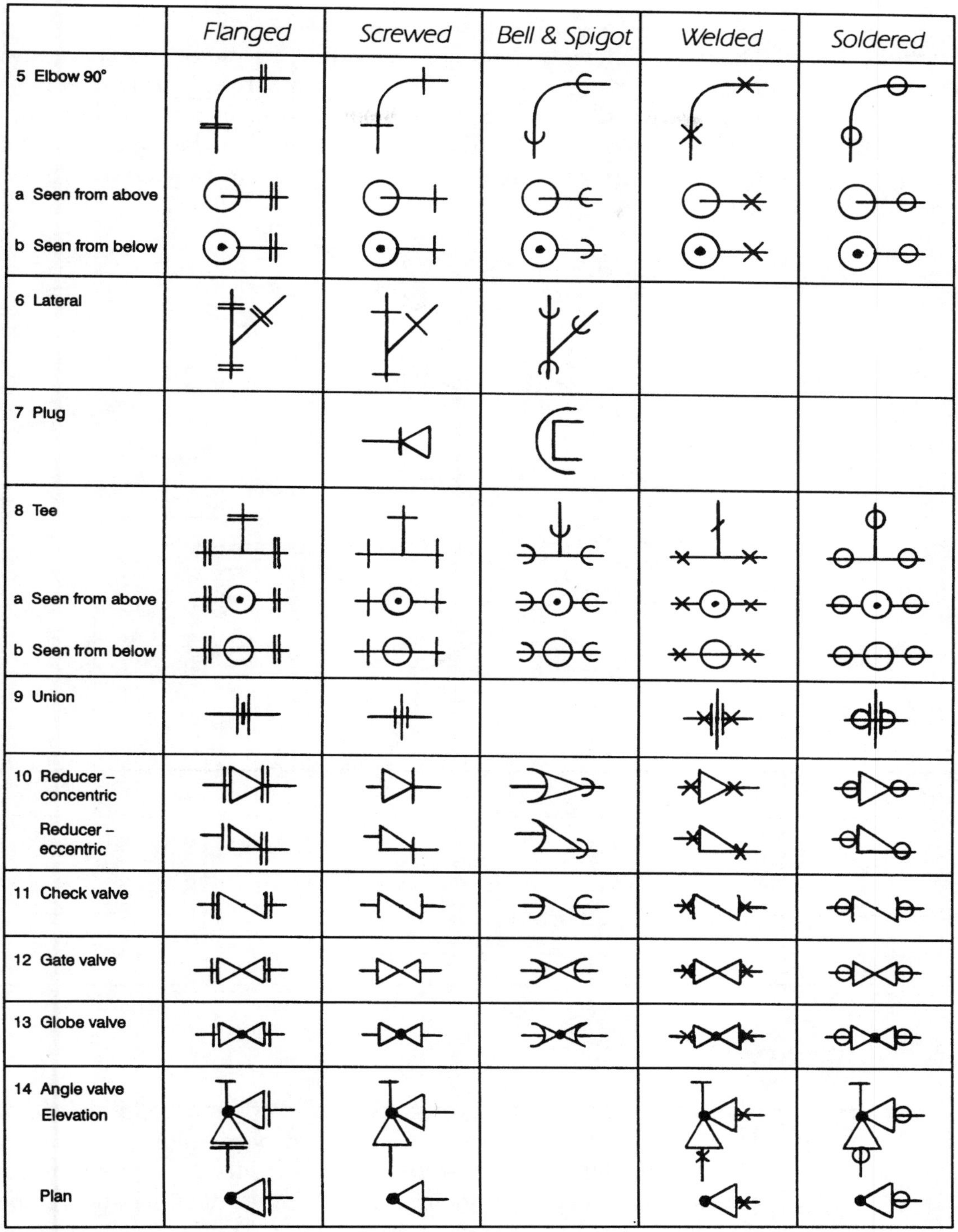
Flanged
Screwed
Bell & Spigot
Welded
Soldered
5 Elbow 90°
a Seen from above
b Seen from below
6 Lateral
7 Plug
8 Tee
a Seen from above
b Seen from below
9 Union
10 Reducer – concentric
Reducer – eccentric
11 Check valve
12 Gate valve
13 Globe valve
14 Angle valve
Elevation
Plan

Activity 1

Draw the following diagram, single-line, in good proportion, and name the fittings clearly. The existing diagram is drawn isometrically with the fittings listed below. Draw, in first-angle orthographic projection, the front view and the top view. Number and label the fittings correctly. Dimensions are in centimetres. Scale 1:1.

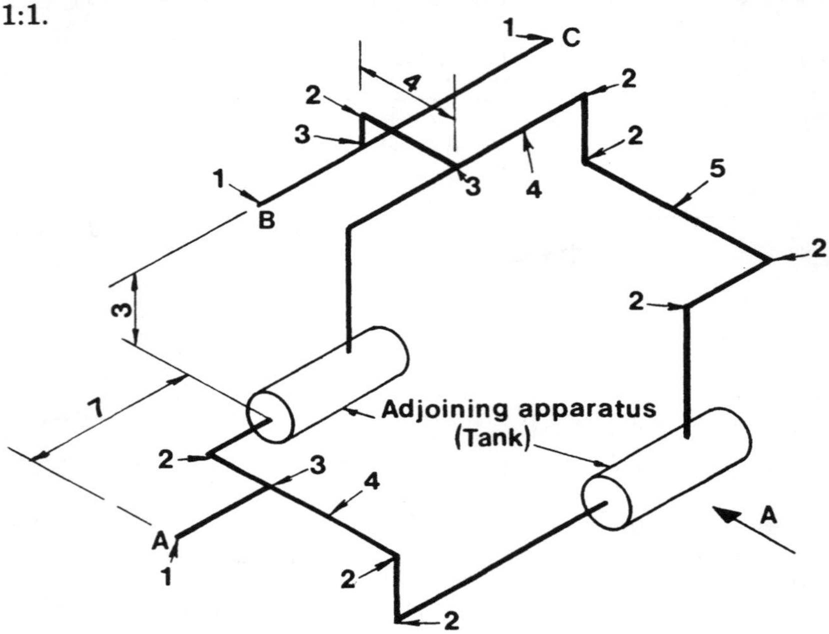

Fig. 11.4 Isometric projection

Listed drawing symbols

1 Flange	**2** Elbow – 90°C
3 Tee	**4** Gate Valve
5 Pipe Line	

6 Summary

This unit has introduced you to the world of piping. Piping is used in water plants, chemical plants and also gas plants. It is part of your syllabus and therefore a good knowledge of piping is expected. Diagrams and symbols for the fittings must be understood clearly so that you can interpret a chemical engineering drawing for a chemical plant. Although it is not necessary to memorise the symbols, you must be able to apply the principles correctly.

Self-evaluation

1 What was the only important fluid that was conveyed from place to place in pipes until recently?
2 Name the different kinds of pipes.
3 Name the three general classes of fittings.
4 Name three other methods for joining of pipes.
5 Name the different types of pipe drawings.

Answers

Activity 1

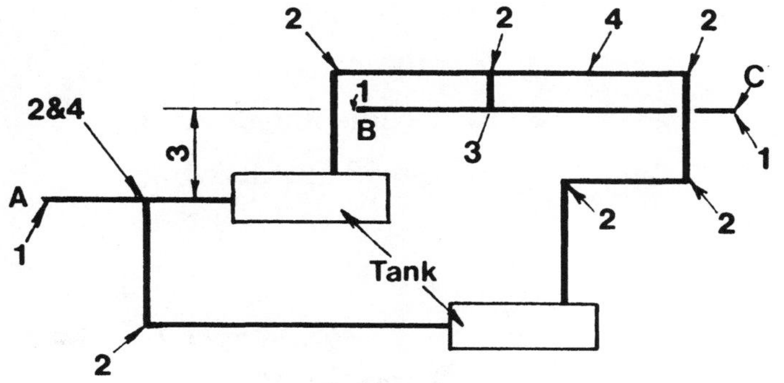

Fig. 11.5 Front view (Direction arrow A)

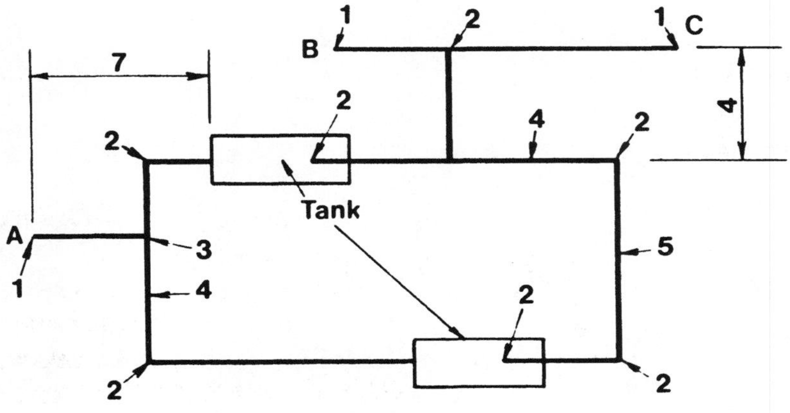

Fig. 11.6 Top view

Self-evaluation

1 Water

2 a Steel and wrought iron piping

  b Cast iron pipes

  c Seamless brass or copper pipes

  d Copper tubing

  e Plastic pipes

3 a Screwed

  b Welded

  c Flanged

4 a Soldering

  b Brazing

  c Gluing

5 a Double line drawings

  b Single line drawings

Appendix
Terms, abbreviations and symbols

General terms, abbreviations and symbols			
Across flats	AF	**Hexagon head**	HEX HD
Assembly	ASSY	**Hydraulic**	HYD
British Association	BA*	**Included**	INC
British Standard	BS	**Insulated or insulation**	INSUL
British Standard Whitworth	BWhit*	**Internal diameter**	ID
British Standard fine	BSFine*	**International Organisation for Standardization**	ISO
British Standard cycle	BSCycle*	**Left-hand**	LH
Centre line	C_L or CL	**Long**	LG
Centres	CRS	**Machine**	MC
Chamfer	CHAM	**Machined**	MCD
Cheese head	CH HD	**Material**	MATL
Concentric	CONC	**Maximum**	MAX.
Countersunk	CSK	**Maximum material condition**	MMC
Countersunk head	CSK HD	**Metric thread nominal diameter of**	M
Counterbore	CBORE	**Micrometre**	μm
Cylinder or cylindrical	CYL	**Minimum**	MIN.
Datum system	DATUM	**Minute (of angle)**	. . .′
Degree	°	**Module**	MOD
Degree Celsius	°C	**Nominal**	NOM.
Diameter (preceding a dimension)	Ø	**Not to scale**	NTS.
Diameter (in a note)	DIA	**Number**	No.
Drawing	DRG	**Outside diameter**	OD
Equal	= or EQ	**Parallel**	PAR
Equally spaced	EQUI SP	**Pattern number**	PATT NO.
Figure	FIG.	**Perpendicular**	PERP
Galvanize		**Pitch circle diameter**	PCD
Hardness –		**Plate**	PL
Brinell	HB	**Pneumatic**	PNEU
Rockwell		**Radian**	rad
A scale	HRA	**Radius**	R
B scale	HRB	**Required**	REQD
C scale	HRC	**Revolutions per minute**	r/min or r.p.m.
D scale	HRD	**Right-hand**	RH
E scale	HRE	**Round head**	RD HD
Vickers	HV		
Hexagon	HEX		

* See relevant BS

General terms, abbreviations and symbols

Screwed	SCR	**Spotface**	SFACE
Second (of angle)	. . . ″	**Square**	SQ or □
Sheet	SHT	**Standard**	STD
Sheet, when preceding the name of a material	SH	**Taper, on diameter or width**	
Sketch	SK	**Tolerance**	TOL
Slope		**Typical**	TYP
South African Bureau of Standards	SABS	**Undercut**	UCUT
Specification	SPEC	**Unified threads**	UNF*
Spherical	SPH	**Volume**	VOL

Abbreviation of terms relating to dimensions and tolerances

Basic dimension	BASIC	**Position**	POSN TOL
Datum **Datum system** **Datum dimension**	DATUM	**Roundness**	RD TOL
		Straightness	STR TOL
Full indicator movement	FIM	**Symmetry**	SYM TOL
Tolerance	**TOL**	**Tolerance zone (profiles)**	TOL ZONE
Angularity	ANG TOL	**True position, or true profile, dimension in conjunction with positional or profile tolerances**	TP
Concentricity	CONC TOL		
Cylindricity	CYL TOL		
Flatness	FLAT TOL		
Parallelism	PAR TOL		